I0057445

General Yang–Mills Symmetry

From Quark Confinement to an Antimatter Half-Universe

ADVANCED SERIES ON THEORETICAL PHYSICAL SCIENCE
A Collaboration between World Scientific and Institute of Theoretical Physics

ISSN: 1793-1495

Series Editor: Yue-Liang Wu
(Institute of Theoretical Physics, Chinese Academy of Sciences, China)

Published

Vol. 14: *General Yang–Mills Symmetry:*
From Quark Confinement to an Antimatter Half-Universe
by Jong-Ping Hsu & Leonardo Hsu

Vol. 13: *Foundations of the Hyperunified Field Theory*
by Yue-Liang Wu

Vol. 12: *Space–Time, Yang–Mills Gravity, and Dynamics of Cosmic Expansion:*
How Quantum Yang–Mills Gravity in the Super-Macroscopic Limit Leads to an Effective $G_{\mu\nu}(t)$ and New Perspectives on Hubble's Law, the Cosmic Redshift and Dark Energy
by Jong-Ping Hsu & Leonardo Hsu

Vol. 11: *Space-Time Symmetry and Quantum Yang–Mills Gravity:*
How Space-Time Translational Gauge Symmetry Enables the Unification of Gravity with Other Forces
by Jong-Ping Hsu & Leonardo Hsu

Vol. 10: *A Broader View of Relativity: General Implications of Lorentz and Poincaré Invariance*
by Jong-Ping Hsu & Leonardo Hsu

Vol. 9: *100 Years of Gravity and Accelerated Frames: The Deepest Insights of Einstein and Yang–Mills*
edited by Jong-Ping Hsu & Dana Fine

Vol. 8: *Lorentz and Poincaré Invariance: 100 Years of Relativity*
by Jong-Ping Hsu & Yuan-Zhong Zhang

Vol. 7: *Einstein's Relativity and Beyond: New Symmetry Approaches*
by Jong-Ping Hsu

Vol. 6: *Differential Geometry for Physicists*
by Bo-Yu Hou & Bo-Yuan Hou

More information on this series can also be found at
https://www.worldscientific.com/series/astps

Advanced Series on Theoretical Physical Science   Volume **14**

General Yang–Mills Symmetry

From Quark Confinement to an Antimatter Half-Universe

Jong-Ping Hsu
University of Massachusetts Dartmouth, USA

Leonardo Hsu
Santa Rosa Junior College, USA

World Scientific

NEW JERSEY · LONDON · SINGAPORE · BEIJING · SHANGHAI · HONG KONG · TAIPEI · CHENNAI · TOKYO

Published by

World Scientific Publishing Co. Pte. Ltd.

5 Toh Tuck Link, Singapore 596224

USA office: 27 Warren Street, Suite 401-402, Hackensack, NJ 07601

UK office: 57 Shelton Street, Covent Garden, London WC2H 9HE

Library of Congress Control Number: 2023902565

British Library Cataloguing-in-Publication Data
A catalogue record for this book is available from the British Library.

Advanced Series on Theoretical Physical Science — Vol. 14
GENERAL YANG–MILLS SYMMETRY
From Quark Confinement to an Antimatter Half-Universe

Copyright © 2024 by World Scientific Publishing Co. Pte. Ltd.

All rights reserved. This book, or parts thereof, may not be reproduced in any form or by any means, electronic or mechanical, including photocopying, recording or any information storage and retrieval system now known or to be invented, without written permission from the publisher.

For photocopying of material in this volume, please pay a copying fee through the Copyright Clearance Center, Inc., 222 Rosewood Drive, Danvers, MA 01923, USA. In this case permission to photocopy is not required from the publisher.

ISBN 978-981-122-290-0 (hardcover)
ISBN 978-981-122-291-7 (ebook for institutions)
ISBN 978-981-122-292-4 (ebook for individuals)

For any available supplementary material, please visit
https://www.worldscientific.com/worldscibooks/10.1142/11900#t=suppl

Desk Editor: Carmen Teo Bin Jie

This monograph is dedicated to

T. Y. Wu, S. Okubo, and E. C. G. Sudarshan

Preface

The goal of this monograph is to provide new perspectives and explanations for current problems in physics, including the late-time cosmic acceleration, flat galactic rotation curves, a permanent quark-confining potential, and the dearth of antimatter. Rather than proposing radically new particles or phenomena that require major modifications to the existing framework of physics, we aim to accomplish this based on established principles of particle physics and gauge fields. It is gratifying that quantum Yang-Mills gravity can be formulated in flat spacetime with external space-time gauge symmetry. This allows us to develop a logically consistent Broad Particle-Cosmology that includes

(i) general Yang-Mills symmetry,

(ii) particle physics with quantum Yang-Mills gravity, and

(iii) inertial and non-inertial frames of reference with space-time transformations among them. Importantly, the space and time coordinates of those frames have operational meanings.

We hope this monograph will be useful to students and researchers interested in a unified picture of the universe from the microscopic to the super-macroscopic scales.

Our starting point is the generalization of the gauge symmetry principle. General Yang-Mills symmetry utilizes a vector gauge function $\omega_\mu(x)$ rather than the arbitrary scalar gauge function $\omega(x)$, where $x \equiv (ct, \mathbf{r})$. Conventional gauge symmetry thus becomes a special case of general Yang-Mills symmetry when $\omega_\mu(x)$ is the space-time derivative of a scalar function, $\omega_\mu(x) = \partial_\mu \omega(x)$.

As a result of this change, not only is the phase factor a non-integral phase factor, it must be a Hamilton's characteristic function. Furthermore, we now have a new strange field, called a 'phase field,' which obeys fourth-order field equations. All hell breaks loose as the phase field quantum, which we call a 'confion,' turns out to be massless and have zero energy-momentum! Although they are thus not directly observable, they may be the first instance of a Wigner particle of the third class, where Wigner's classification is based on the unitary irreducible representations of the Poincaré algebra.

The Feynman propagator for a confion resembles the square of a photon propagator. A virtual confion can be exchanged between quarks to produce, say, a static linear potential, similar to a virtual photon in QED in some sense. However incomprehensible such properties may be, we provide as much physical and mathematical reasoning as we can, along with possible observational consequences that can be used to test the viability of this idea. We believe that the idea of the confion is too important to be ignored, especially as a logical consequence of general Yang-Mills symmetry.

In the microscopic world with general color SU_3 symmetry, the static confion field equations lead to dual linear and Coulomb-like potentials, which are consistent with the measured spectrum of charmonium. In the super-macroscopic world with general baryonic gauge symmetry, extremely small but ubiquitous baryonic charges can produce a super long-range repulsive force that may be responsible for the observed late-time cosmic acceleration.

We propose that CPT invariance in particle physics implies a maximum particle-antiparticle symmetry and a Big Jets event, rather than a Big Bang, for the beginning of the universe. The Big Jets beginning can be pictured as the creation of two gigantic fire balls moving away from each other. One is dominated by particles and the other by antiparticles, and each is a Big Bang in its own right. Now, billions of years later, both fireballs have cooled to form a 3K blackbody (our observable universe) and a 3K anti-matter blackbody a great distance away. We believe the missing anti-matter, located in the as yet unobserved 3K anti-matter blackbody, might be detected as a very weak hemispheric asymmetry in the cosmic microwave background in the data of the Planck satellite.

General Yang-Mills symmetry has the ability to show us connections between the microscopic quark-confining force and a possible super-macroscopic baryonic force responsible for cosmic acceleration. Similarly, CPT invariance shows us connections between equal particle-antiparticle creation at the microscopic scale in high energy laboratories and a possible creation of matter-antimatter half-universes at the super-macroscopic scale.

Finally, we discuss a model for the unification of all interactions, including gravity, in which all interactions have their own general Yang-Mills symmetry groups. It is not required for all interactions to be characterized by the same coupling constant under the same Lie group. This model reveals that all internal gauge symmetries have an extremely small violation due to their coupling to quantum Yang-Mills gravity.

The present work would not have been possible were it not for the publication of our earlier volume, *Space-Time, Yang-Mills Gravity and Dynamics of Cosmic Expansion* (World Scientific, 2019), which received an encouraging book review from Marek Nowakowski: "The theories presented in the book are mathematically consistent and explained in a pedagogical way. What I like most about the book is the fact that the authors pay attention to the observational aspect of the theory they propose. This is how science should be done and exposed, starting with a mathematical framework and ending with numbers which can be compared with the experiments. The book is both courageous in developing new concepts and ideas, and honest about the numerical consequences." (Mathematical Reviews Clippings, May 2021.) The purpose of this monograph is to push our ideas to the next level, continuing in that same spirit.

We would like to thank Luree Schneider and Leslie Hsu for their loving support in all these years, and Bonnie Hsu for her continuing support throughout our lives.

J. P. Hsu and L. Hsu

Contents

Part II: Symmetry-Unified Quark-Cosmic Model Based On General Yang-Mills Symmetry   91

Chapter 6. A Universal Principle of Interactions for Quarks and Leptons   93

Chapter 7. Finite Fermion Self-Masses and a Non-Propagating Phase Field   105

Chapter 8. Quark Confinement and the Accelerated Cosmic Expansion   117

Supreme Beauty

SYMMETRY

Symmetry, as wide or as narrow as you may define its meaning, is one idea by which man through the ages has tried to comprehend and create order, beauty and perfection. (H. Weyl)

Nature seems to take advantage of the simple mathematical representations of the symmetry laws. When one pauses to consider the elegance and the beautiful perfection of the mathematical reasoning involved and contrast it with the complex and far-reaching physical consequences, a deep sense of respect never fails to develop. (Nobel Laureate, C. N. Yang)

General Yang-Mills symmetry has the ability to show us connections between the microscopic quark-confining force and a possible super-macroscopic baryonic force responsible for cosmic acceleration. Similarly, CPT invariance shows us connections between equal particle-antiparticle creation at the microscopic scale in high energy laboratories and a possible creation of matter-antimatter half-universes at the super-macroscopic scale. (Preface, Authors)

0

Overview

0-1. Symmetry: The simplest and most far-reaching concept in foundational physics

Symmetry in foundational physics is simple but extremely intriguing; it is arguably the most far-reaching concept in physics. In particular, general gauge symmetry is the underpinning of our understanding of all forces in nature from quark confinement at the microscopic scale ($\approx 10^{-15}m$) to the accelerated cosmic expansion at the super-macroscopic scale ($\geq 10^{23}m$).

One may wonder: What is symmetry in physics? Feynman's interpretation of mathematician-physicist H. Weyl's definition of symmetry is as follows: "a thing is symmetrical if one can subject it to a certain operation and it appears exactly the same after the operation." [1]

If one takes a sphere as 'a thing,' then one can move it to another location or rotate it by an arbitrary angle. After one has finished, the sphere looks exactly the same as it did before. In this case, one says that the sphere is symmetrical (or invariant) under a translation in space by an arbitrary displacement or under rotation by an arbitrary angle.

Now suppose a 'thing' means the equation of a circle with radius a on the x-y plane, or the law for the propagation of light,

$$x^2 + y^2 = a^2; \tag{0.1}$$

$$x^2 + y^2 + z^2 = c^2 t^2,$$

then what one can do to is to rotate coordinates (or make a Lorentz transformation of space-time coordinates),

$$x = x' cos\theta' + y' sin\theta', \qquad y = -x' sin\theta' + y' cos\theta'; \tag{0.2}$$

$$x = \gamma(x' + \beta ct'), \quad y = y', \quad z = z', \quad ct = \gamma(ct' + \beta x'),$$

$$\gamma = 1/\sqrt{1 - \beta^2}, \quad \beta = V/c.$$

The first change of coordinates may be pictured as rotating the two-dimensional coordinate axes by an arbitrary angle θ'. A Lorentz transformation is a change of space-time coordinates from one inertial frame $F(ct, x, y, z)$ to another inertial frame $F'(ct', x', y', z')$. By substituting

(0.2) into (0.1), one obtains another equation of the same form or the law for the propagation of light as observed in another frame $F'(ct', x', y', z')$,

$$x'^2 + y'^2 = r^2; \qquad (0.3)$$

$$x'^2 + y'^2 + z'^2 = c^2 t'^2.$$

The law for the propagation of light as observed from two inertial frames $F(ct, x, y, z)$ and $F'(ct', x', y', z')$ have the same form. This is the heart of the four-dimensional symmetry of space-time.

One can say that Eq. (0.1) is symmetric (or invariant) under a rotation of the coordinates. This is intuitively clear because the circle is not affected by the action of rotating the coordinate axes. We see that if an equation has a geometric meaning, the symmetry of the equation means that a certain geometric property does not change. Here, Eqs. (0.1) and (0.3) show that the length of a rod does not change under a rotation of coordinates by an arbitrary angle. In the second example, the equation describes the propagation of light with time. In this case, we are dealing with the law of propagation of light that involves time and hence, is not just a simple static geometric object or a picture.

From a microscopic viewpoint, roughly 98% of atoms in the universe are hydrogen and helium, which have spherically symmetric ground states. The physical world is full of symmetric spheres, from cosmic structures such as stars to tiny liquid droplets in the air. These shapes are consequences of the basic laws of physics with spherical symmetry. Nevertheless, these symmetric basic laws also lead to innumerable non-spherically symmetric structures such as crystals. All these phenomena reveal the profound and far-reaching consequences of symmetries in physical laws.

A convincing reason for symmetry to play an important role in physics is that it leads to fundamental conservation laws.[a] This connection is revealed through Noether's theorem,[2] which states that if the Lagrangian (or the action) of a physical system is symmetric or invariant under a transformation, then there is a corresponding conservation law for the system. Such a symmetry associated with a conservation law in the Lagrangian is called a 'geometric symmetry.'

A far more significant role played by the symmetry of a Lagrangian is related to the dynamics of interactions in physics. For example, the laws governing the electromagnetic properties of a physical system, such as

---

[a]If a symmetry does not lead to a conservation law, we may conjecture that it may not be a truly fundamental symmetry in physics and less useful in quantum field theory, since all established conservation laws are associated with symmetries.

Maxwell equations, can be derived from a Lagrangian L_{em} that involves the electromagnetic vector potential A_μ and has a special local gauge symmetry associated with the U_1 group or the U_1 gauge symmetry. The procedure is as follows: First one defines the U_1 gauge transformations for the electromagnetic potential $A_\mu(x)$ with an arbitrary scalar function of space-time $\Lambda(x)$

$$A_\mu(x) \to A'_\mu(x) = A_\mu(x) + \partial_\mu \Lambda(x), \qquad c = \hbar = 1, \quad x = x^\nu, \qquad (0.4)$$

which are analogous to the coordinate transformations (0.2). The U_1 gauge curvature $F_{\mu\nu}(x)$ is defined by the commutator of the gauge covariant derivative $\Delta_\mu = \partial_\mu - ieA_\mu$,

$$[\Delta_\mu, \Delta_\nu] = -ieF_{\mu\nu}, \qquad F_{\mu\nu}(x) = \partial_\mu A_\nu(x) - \partial_\nu A_\mu(x). \qquad (0.5)$$

The curvature tensor $F_{\mu\nu}(x)$ is invariant under the gauge transformation (0.4),

$$F_{\mu\nu}(x) \to F'_{\mu\nu}(x) = \partial_\mu A'_\nu(x) - \partial_\nu A'_\mu(x) = F_{\mu\nu}(x). \qquad (0.6)$$

This invariant tensor $F'_{\mu\nu}(x) = F_{\mu\nu}(x)$ plays a key role in the derivation of Maxwell's equations. One can construct a simple U_1 gauge invariant Lagrangian L_{em} that is quadratic in the gauge curvature. Under the gauge transformations (0.4), we have

$$L_{em} = -\frac{1}{4}F_{\mu\nu}F^{\mu\nu} \quad \to \quad L'_{em} = -\frac{1}{4}F'_{\mu\nu}F'^{\mu\nu} = L_{em}. \qquad (0.7)$$

Thus, the Lagrangian L_{em} is U_1 invariant (and also Lorentz invariant). We note that Eq. (0.6) is analogous to (0.1) and (0.3) in the previous discussions

The U_1 gauge invariant L_{em} and $F_{\mu\nu}(x)$[b] lead to Maxwell equations, which predict the propagation of electromagnetic waves in vacuum. Similarly, based on U_1 gauge symmetry, one can derive the basic equation of motion for a classical and quantum electron. These are really intriguing, wonderful and far-reaching consequences in physics for our understanding of nature.

To stress the crucial role played by symmetries in the basic laws of physics, Wigner observed that symmetries can be divided into two classes:

---

[b]The Lagrangian L_{em} leads to Ampere's law and Gauss' law. One also has to use the identity associated with $F_{\mu\nu}$, namely, $\partial_\lambda F_{\mu\nu} + \partial_\mu F_{\nu\lambda} + \partial_\nu F_{\lambda\mu} = 0$, to obtain Faraday's law and the absence of the magnetic monopole $\nabla \cdot \mathbf{B} = 0$.

Class 1: Geometric Symmetries. These lead to conservation laws only, and are not directly related to the dynamics of interactions.

Class 2: Dynamic Symmetries. These dictate the dynamics of interactions in physics, such as the electroweak and the strong (or quark) interactions, in addition to having corresponding conservation laws.

The most fundamental geometric symmetry in physics is the 4-dimensional space-time symmetry with the Poincaré group. It includes the Lorentz group as a subgroup and leads to the conservation of energy-momentum and of angular momentum. Wigner stated that the geometrical principle of invariance was 'recognized by Poincaré first, and I like to call the group formed by these invariables the Poincaré group.'[3,c]

The dynamic symmetries have far-reaching consequences and their importance cannot be emphasized strongly enough in foundational physics. It now appears clear that dynamical symmetries are local gauge symmetries. The generators of the dynamical groups, such as $SU_2 \times U_1$ and color SU_{3c} for the electroweak theory and quantum chromodynamics, respectively, have constant matrix representations. The gauge transformations of these dynamical groups correspond to phase changes. These are key properties of conventional gauge theories with internal symmetry groups. Analogously, Yang-Mills gravity is based on the postulate that the 'external' space-time translational group is also a dynamical group, even though its group generators do not have dimensionless constant matrix representations. Though unconventional, this property is just right for the gravitational interactions because the gravitational force is always attractive. (See (0.8)–(0.11) below.) Furthermore, the gravitational coupling constant has the dimension of length in natural units and is naturally related to the Newtonian constant.

0-2. The principle of general Yang-Mills symmetry

In this monograph, we introduce general Yang-Mills symmetry, which is a universal dynamic symmetry. It is based on vector gauge functions $\omega_\lambda(x)$ in gauge transformations, involving Hamilton's characteristic phase function.[4,5]

Within the total symmetry-unified model that we discuss below, which can accommodate all known and perhaps some as-yet-undiscovered forces that we will also discuss, the universal principle of dynamic symmetry is postulated as follows:

---

[c]The name 'Poincaré group' was also used in a set of lecture notes by A. S. Wightman around 1964 (as related by R. F. Streater).

All dynamic symmetries with internal and external gauge groups are dictated by general Yang-Mills symmetries with vector gauge functions $\omega_\mu(x)$ in the gauge transformations. The conventional gauge symmetries are then special cases of general Yang-Mills symmetries in which the vector functions can be expressed as the space-time derivative of scalar functions, e.g., $\omega_\mu(x) = \partial_\mu \Lambda(x)$.

0-3. The nature of forces and why is the gravitational force always attractive?

One of the oldest questions in physics is why the gravitational interaction is always attractive and never repulsive while other forces, such as the electromagnetic force (and others that we will discuss, that may not have been recognized yet) appear to be capable of being both attractive and repulsive. Quantum Yang-Mills gravity can answer this question. The gauge covariant derivative δ_μ through which quantum Yang-Mills gravity couples to all other physical fields, is given by

$$\delta_\mu = \partial_\mu - ig\phi_\mu^\nu p_\nu - ieA_\mu + ... = \partial_\mu + g\phi_\mu^\nu \partial_\nu - ieA_\mu + ... \quad (0.8)$$

$$(\delta_\mu)^* = (\partial_\mu - ig\phi_\mu^\nu p_\nu - ieA_\mu + ...)^* = \partial_\mu + g\phi_\mu^\nu \partial_\nu + ieA_\mu + ... \quad (0.9)$$

where $c = \hbar = 1$ and $p_\mu = i\partial_\mu$ is the generator of the space-time translation T_4 group. The gauge covariant derivative δ_μ and its complex conjugate $(\delta_\mu)^*$ appear in the wave equations of the electron (with charge e) and the positron (with charge $-e$), respectively. Since the electric force between two charged particles can be considered to be a consequence of the exchange of a virtual photon,[d] the electric force between two electrons and between an electron and a positron are respectively given by the third terms in (0.8) and (0.9), i.e.,

$$(-ie) \times (-ie) = -e^2, \quad repulsive, \quad (0.10)$$

$$(-ie) \times (ie) = +e^2 \quad attractive.$$

Thus, the theory predicts the experimentally observed attractive and repulsive electric forces, which are due to the presence of i in the electromagnetic

---

[d]This can be pictured in a Feynman diagram as two vertices connected by a photon propagator in quantum electrodynamics.

coupling. In contrast, the gravitational force between two electrons and between an electron and a positron are respectively given by the second terms in (0.8) and (0.9), giving us an attractive gravitational force in both cases.

$$(g) \times (g) = +g^2, \quad attractive, \tag{0.11}$$

$$(g) \times (g) = +g^2, \quad attractive.$$

The key reason is that the gravitational coupling terms in (0.8) and (0.9) do not involve i, in contrast with the coupling terms for the electromagnetic force. We note that the qualitative results for forces in (0.10) and (0.11) are independent of the signs of the coupling constants e and g. Furthermore, the gravitational coupling constant g in (0.8) and (0.9) has the dimension of length, in contrast to all other coupling constants of fields associated with internal gauge groups. Furthermore, g^2 is related to the Newtonian constant G by $g^2 = 8\pi G$.[6,7] Such an important qualitative result (0.11) at the microscopic quantum level indicates that the space-time translation gauge group of Yang-Mills gravity is just right for gravity.

0-4.   A total symmetry-unified model

Based on the framework of flat space-time and the experimentally established conservation laws, we propose a total symmetry-unified model that can encompass all known interactions. It includes the following total symmetry groups:

$$G_{tot} = (T_4)_{YMgravity} \times [SU_{3c}]_{color} \times (SU_2 \times U_1) \times U_{1b} \times U_{1\ell}. \tag{0.12}$$

The general covariant derivative for the groups G_{tot} takes the form[e]

$$\delta_\mu = \partial_\mu + g\phi_\mu^\nu \partial_\nu + ig_c H_\mu^a \frac{\lambda^a}{2} + ifW_\mu^m t^m + if'U_\mu t_o + ig_b B_\mu + ig_\ell L_\mu, \tag{0.13}$$

where $(\lambda^a/2), a = 1, 2, ...8$, are the SU_3 group generators; t^m, m = 1, 2, 3, are weak isospins (or SU_2 generators in general) and t_o is the weak hypercharge (or the U_1 generator).[8] The T_4 generator, $p_\mu = i\partial/\partial^\mu$, differs from the generators of all other internal gauge groups, which have constant matrix representations.

At first glance, it may seem as if the total symmetry-unified model simply juxtaposes the various gauge fields and group generators without any true unification. However, we shall argue that this level of unification

---

[e]When one applies this δ_μ to a quark wave function one should set the leptonic charge g_ℓ to zero because the quark does not carry the leptonic charge.

is probably all that can be done to understand the physical world as we know it for the following reasons:

(A) Mathematically, it appears unlikely that one can further unify these external and internal gauge symmetries in (0.12) into a single Lie group and a single coupling constant, as expected by the conventional view of unification.

(B) The usual asymptotic freedom and running coupling constants in non-Abelian gauge theories are, strictly speaking, only approximately true. This is because coupling to the universal gravitational interaction upsets the renormalizability of all Abelian and non-Abelian gauge theories with internal symmetry groups.

The total symmetry-unified model with (0.12) reveals that all internal gauge symmetries have an extremely small violation because all physical fields couple to quantum Yang-Mills gravity. For example, the electric charge is predicted to be modified by the gravitational potential of the Earth at a level of 1 part in $\approx 10^9$. In chapter 10 we discuss some experimental implications of this violation. In Appendix B, we discuss a test of quantum Yang-Mills gravity with modified Josephson current-phase relations and with a new gauge invariant phase equation for Cooper pairs. The new phase equation is consistent with flux quantization in a superconducting ring and with the Meissner effect. The model with (0.12) further reveals that *the space-time translation T_4 symmetry associated with gravity (which also implies the conservation of energy-momentum) is the only truly exact gauge symmetry in nature.*

0-5.   A big picture of the universe

Broader Particle-Cosmology is the fundamental framework by which the universe, at all scales, can be understood. It is based on three ingredients:

(A) general Yang-Mills symmetries, which include the conventional gauge symmetries as special cases,

(B) particle physics with quantum Yang-Mills gravity in flat space-time with the Poincaré group,

(C) inertial and non-inertial frames of reference with space-time transformations among their coordinates, which have operational meanings.[9]

In light of the well-established CPT theorem[f] in particle physics, the

---

[f] A local quantum field theory is CPT invariant if it is Lorentz invariant and satisfies the spin-statistics relation for quantization of fields. Here, C (charge conjugate) denotes changing the sign of a charge, P(parity) denotes space inversion $\mathbf{r} \to -\mathbf{r}$, and T denotes

absence of antimatter in the observable portion of the universe suggests that the Big Bang model may not be the best framework for understanding the physical universe. We propose a simple Big Jets model for the birth of the universe.[5,10] This model[11] suggests that the birth of the universe can be pictured as the creation of two diametrically opposed jets, similar to the type of phenomena one might encounter in a particle collision in a high energy physics laboratory.

Each jet can be modeled as two giant fire balls moving away from each other, each similar to a Big Bang in its own right. Although equal numbers of particles and anti-particles were created in the Big Jets event, one of the two fire balls had an excess of particles, and the other, of anti-particles. The formation of hadrons and anti-hadrons by quarks and antiquarks, the annihilation of particles and antiparticles, and natural decay processes eventually led to one fire ball being dominated by particles, and the other by anti-particles. From the vantage point of an observer in each fire ball, the evolution of each would be similar to the general features of a hot Big Bang. Over the course of billions of years, each fire ball cooled to form a gigantic cluster of matter (or anti-matter) galaxies. Each can also be modeled as a blackbody with a temperature of roughly 3K. Thus, the Big Jets model is consistent with all the experimentally established conservation laws of electric charge, baryon number, electron-lepton number, etc.[g]

References

1. H. Weyl, *Symmetry* (Princeton Univ. Press, 1952); R. P. Feynman, R. B. Leighton, M. Sands, *The Feynman Lectures on Physics*, (Addison-Wesley, 1965), Vol. 1, Ch. 11, p. 1.
2. E. Noether, Goett. Nachr. 235 (1918). English translation is online: Google search "M. A. Tavel, Noether's paper." (Downloaded on Dec. 10, 2022.)
3. E. Wigner, Proc. Nat'l Acad. Sci. 51, No. 5 (May, 1964) and 'Symmetries and Reflections, Scientific Essays' (MIT Press, 1967) pp. 18–19.
4. W. Yourgrau, S. Mandelstam, 'Variational Principles in Dynamics and Quantum Theory', (3rd ed, 1979, Dover) p. 50.
5. J. P. Hsu, Chin. Phys. C **41**, 015101 (2017).
6. J. P. Hsu, Int. J. Mod. Phys. A **21** 5119 (2006); arXiv: 1102.2010 [gr-qc].
7. J. P. Hsu, Int. J. Mod. Phys. A **24** 5217 (2009).
8. K. Huang, *Quarks, Leptons and Gauge Fields* (World Scientific, 1982), pp. 103–121 and pp. 241–242.

---

time reversal $t \to -t$.[12]

[g]Here, the gravitational effects for their violations are extremely small and can be neglected.

9. J. P. Hsu and L. Hsu, *Space-Time, Yang-Mills Gravity, and Dynamics of Cosmic Expansion*, (World Scientific, 2013) pp. 28–30.
10. For interesting discussions on cosmology, see T. Rothman and E. C. G. Sudarshan, *Doubt and Certainty* (With ill. by Shannon K. Comins, Helix Book, Reading, Mass. 1998) pp. 214–264.
11. J. P. Hsu, L. Hsu, D. Katz, Mod. Phys. Lett. A **33**, 1850116 (2018).
12. T. D. Lee, *Particle Physics and Introduction to Field Theory* (Harwood Academic Publishers, 1981) pp. 746–752.

About the Authors

Jong-Ping Hsu received his B.S. from the National Taiwan University and M.S. from the National Tsing Hua University. He earned his Ph.D. in 1969 studying particle physics at the University of Rochester (New York) with Professor S. Okubo. He has done research at McGill University, Rutgers, the University of Texas at Austin, the Marshall Space Flight Center, NASA, and the University of Massachusetts, Dartmouth. He has been a visiting scientist at Brown, MIT, Taiwan University, Beijing Normal University, and the Academy of Science, China. His research is concentrated in the areas of gauge field theories, Yang-Mills gravity, a total-unified model of interactions and broad views of 4-dimensional symmetry. He has published more than 175 papers and articles, and two books, "A Broader View of Relativity, General Implications of Lorentz and Poincaré Invariance" (with L. Hsu, World Scientific, 2006) and "Lorentz and Poincaré Invariance" (with Y. Z. Zhang, World Scientific, 2001). He has also co-edited four conference proceedings and the book "100 Years of Gravity and Accelerated Frames — The Deepest Insights of Einstein and Yang-Mills" (with D. Fine, World Scientific, 2005). He is currently a chancellor professor and the Director of the Jing Shin Research Fund at the University of Massachusetts, Dartmouth.

Leon Hsu earned his Ph.D. studying semiconductor physics at the University of California, Berkeley. After doing a postdoc at the Center for Innovation in Learning at Carnegie Mellon University, he joined the faculty of the University of Minnesota, Twin Cities in 2000, before moving to Santa Rosa Junior College in 2017. His research has included both experimental and theoretical work. His primary interest is in how students learn to solve problems in introductory physics courses and the attitudes and beliefs of students regarding physics and the learning of physics. He has also studied transport processes in semiconductors, as well as implications of the principle of relativity and different systems of units for different relativity theories. He was a co-editor of the "JingShin Theoretical Physics Symposium in Honor of Professor Ta-You Wu" (World Scientific, 1998). His work has been supported by the Potz Science Fund and by the National Science Foundation.

Part I

Broader Particle-Cosmology:

New Perspectives On Dark Matter,

Dark Energy, And The Missing Anti-Matter

1

Underpinnings

1-1. Broader Particle-Cosmology in flat space-time

Broader Particle-Cosmology denotes a framework for the fundamental interactions of particle physics using Yang-Mills gravity in flat space-time, together with the cosmological principle embedded in Lagrangian dynamics. Although much of the framework of Broader Particle-Cosmology has been published elsewhere, in the interest of keeping this book self-contained, we review the foundational ideas on which this framework is based, which are:

A. Inertial and non-inertial frames with operationally defined space and time coordinates.

B. Quantum and Classical Yang-Mills gravity based on inertial frames in flat space-time.

C. General Yang-Mills Symmetry, a generalization of conventional gauge symmetry that involves Lorentz vector gauge functions.

Our goal is to gain a new understanding of physical phenomena at microscopic, macroscopic and super-macroscopic scales while staying as close to established particle physics as possible.[1] The established basic dynamical laws of modern particle physics are based on gauge symmetries, including unified electroweak theory and quantum chromodynamics. As mentioned previously, each of the foundational ideas above is a generalization of conventional physics that includes the conventional physics as a limiting or special case.

Since this framework is based on flat space-time, its physical implications are expressed using operationally-defined space-time coordinates in inertial and non-inertial frames i.e., its results are as simple to interpret as those in special relativity and much more straight-forward to interpret than the general coordinates of General Relativity. In fact, most discussions can be conducted within the simple well-defined space-time coordinates of inertial frames. A few phenomena, such as redshifts caused by the accelerated motion of receding galaxies and invariant blackbody spectra in inertial and non-inertial frames will be derived with the help of the coordinate transformations for accelerated frames, as we discuss below (and elsewhere[2]). In this chapter, we will briefly discuss the three ideas above and include references to more detailed discussions and derivations.

1-2. The principle of limiting continuation of physical laws in inertial and non-inertial frames

It is worth examining some of the supporting structures of particle physics that are often taken for granted, in particular four-dimensional space-time and the inertial frames of reference that must be part of any physical theory. Traditionally, physical theories have all been formulated in inertial frames even though, strictly speaking, all physically realizable frames are non-inertial due to the long-range gravitational interaction and the accelerated expansion of the universe. Although the inertial frame approximation is adequate for a great number of situations, our understanding of physics should not be restricted to inertial frames since some phenomena, such as the motion of galaxies at super-macroscopic scales fundamentally involve non-inertial frames and physics in accelerated frames is qualitatively different from that in inertial frames.

In order to understand these non-inertial phenomena, we have generalized the familiar Lorentz space-time transformations between inertial frames to a more general set of space-time transformations that encompass both inertial and non-inertial frames within the framework of flat space-time. Because of the complexity of the mathematics, we have limited ourselves to a restricted, but important subset of non-inertial frames, namely those where the acceleration is along a line or where the motion is in a circle with constant speed. Such efforts are not new, of course. However, many of previous attempts began with general relativity,[2–4] and as a result, the obtained space-time coordinate transformations do not reduce smoothly to the Poincaré transformations in the limit of zero acceleration. In view of the comments made by Dyson and Wigner,[3] this may not be surprising.

We have taken a different tack. Because we view it as critical that coordinate transformations for non-inertial frames reduce to the Lorentz transformation in the limit of zero acceleration, our efforts are based on what we call the principle of limiting continuation of physical laws. *This is nothing more than the generalized principle of relativity for space-time transformations for both inertial and non-inertial frames.*[5] Simply put, we postulate:

The laws of physics in a reference frame F_1 with an acceleration $\mathbf{a}_1$ must reduce to those in a reference frame F_2 with an acceleration $\mathbf{a}_2$ in the limit where $\mathbf{a}_1$ approaches $\mathbf{a}_2$.

Since the accelerations $\mathbf{a}_1$ and $\mathbf{a}_2$ are not specified, in the special case that $\mathbf{a}_1 \to \mathbf{a}_2 = 0$, this principle of limiting continuation of physical laws reduces to the principle of relativity in the zero acceleration limit. In this sense, the principle of limiting continuation of physical laws includes 'limiting Poincaré and Lorentz invariance' as a special case, and appears to be a simple generalization of the principle of relativity from inertial frames to non-inertial frames.

Details of this work can be found elsewhere, including discussions of the justification for this approach and explicit calculations showing the consistency between the predictions of our explicit transformations for non-inertial frames and experimental measurements.[5] For the reader's convenience, we now summarize the parts that are critical to our later discussion of particle cosmology and a total unified model with general Yang-Mills symmetry.

Because we aim to create the simplest space-time framework that can accommodate all the laws of physics in a unified manner, we use a unit system that highlights symmetries in nature. Thus, in the explicit transformations that follow, the temporal variable has the dimension of length, the same as the spatial variables, i.e., $x^\mu = (w, x, y, z)$. We use 'w' as the temporal variable to distinguish it from 't' which is commonly associated with a temporal variable expressed in units of 'seconds.' In this system of units, the speed of light $c=1$ does not appear explicitly anywhere. This is consistent with its reduced importance since we do not expect the speed of light to be a universal constant, or even isotropic, in non-inertial frames. More generally, we use Heaviside-Lorentz rationalized units with $\hbar = c = 1$, and $\alpha_e = e^2/(4\pi) = 1/137.036$.

1-3.   Space-time transformations for non-inertial frames

We generalize the explicit coordinate transformations and metric tensors of inertial frames to those for accelerated frames in a series of small steps. First, we make a minimal generalization of the Poincaré transformations[a] (using the principle of limiting continuation) to develop an explicit coordinate transformation between an inertial frame and a non-inertial frame that has a constant linear acceleration (CLA) along the same line as its velocity relative to the inertial frame. Next, we use our results for CLA frames to develop an explicit coordinate transformation between an inertial frame and another frame with an arbitrary linear acceleration (ALA) that is along the same line as its velocity relative to the inertial frame.

---

[a]The Poincaré transformations are the Lorentz transformations with a constant offset in the space-time coordinates.

In all cases, the space-time coordinates retain their traditional operational definitions, unlike the arbitrary coordinates of general relativity.[3]

We begin with the familiar Poincaré transformations between two inertial frames. The law for the propagation of a light signal is given by the invariant law, $ds^2 = dw^2 - dx^2 - dy^2 - dz^2 = 0$, which automatically implies that the speed of light in any inertial frame is a dimensionless universal constant $\beta_c = dr/dw = 1$, where $dr = \sqrt{dx^2 + dy^2 + dz^2}$.

The Poincaré transformations between any two inertial frames $F_I(w_I, x_I, y_I, z_I)$ and $F_i(w, x, y, z)$ are[2]

$$w_I = \gamma_o(w + \beta_o x) + w_o, \qquad x_I = \gamma_o(x + \beta_o w) + x_o, \qquad (1.1)$$

$$y_I = y + y_o, \quad z_I = z + z_o, \qquad \gamma_o = \frac{1}{\sqrt{1 - \beta_o^2}},$$

where the inertial frame $F_i(w, x, y, z)$ moves along the x_I direction of the inertial frame F_I. In the special case, $w_o = x_o = y_o = z_o = 0$, (1.1) simplifies to the Lorentz transformations.

Based on the principle of limiting continuation of physical laws, we now make a minimal generalization from the Poincaré transformations to a set of coordinate transformations between an inertial frame $F_I(w_I, x_I, y_I, z_I)$ and a frame with a constant-linear-acceleration (CLA) $F_{CLA}(w, x, y, z)$, where both the acceleration α_o and velocity β_o of the non-inertial frame point along the x-axis of the inertial frame (they may be parallel or anti-parallel). These transformations are a minimal generalization of (1.1) with the assumption:

$$ds^2 = W^2 dw^2 - dx^2 - dy^2 - dz^2 = dw_I^2 - dx_I^2 - dy_I^2 - dz_I^2, \qquad (1.2)$$

which leads to the following transformations for the coordinate differentials:

$$dw_I = \gamma(W dw + \beta dx), \quad dx_I = \gamma(dx + \beta W dw), \quad dy_I = dy, \quad dz_I = dz; \quad (1.3)$$

$$\gamma = 1/\sqrt{1 - \beta^2},$$

where $W = W(w, x)$ and $\beta = \beta(w)$ are unknown functions to be determined by the integrability conditions of (1.3).

Assuming the usual linear relationship between the velocity β and the time variable w, i.e., $\beta = \alpha_o w + \beta_o$, we obtain $W = \gamma^2(\gamma_o^{-2} + \alpha_o x)$ and[6]

$$w_I = \gamma\beta\left(x + \frac{1}{\alpha_o\gamma_o^2}\right) - \frac{\beta_o}{\alpha_o\gamma_o} + w_o,$$

$$x_I = \gamma\left(x + \frac{1}{\alpha_o\gamma_o^2}\right) - \frac{1}{\alpha_o\gamma_o} + x_o, \quad y_I = y + y_o, \quad z_I = z + z_o; \qquad (1.4)$$

$$\gamma = \frac{1}{\sqrt{1-\beta^2}}, \qquad \gamma_o = \frac{1}{\sqrt{1-\beta_o^2}}, \qquad \beta = \alpha_o w + \beta_o,$$

which satisfies the initial conditions $x_I = x_o$ and $w_I = w_o$ when $w = x = 0$. We call (1.4) the Wu transformations in honor of T. Y. Wu's idea of a kinematical approach to deriving transformations between accelerated frames in a space-time with zero Riemann-Christoffel curvature.[4]

One can verify that the Wu transformations (1.4) reduce to the Poincaré transformations (1.1) in the limit of zero acceleration $\alpha_o \to 0$.[b] This smooth connection with the Lorentz transformation in the limit of zero accelera-tion when $w_o = x_o = y_o = z_o = 0$ in (1.1), is a crucial property of the accelerated transformations not found in other proposed transformations[5] for non-inertial frames.

In the special case, when $\beta_o = 0$ and $w_o = x_o = y_o = z_o = 0$, the Wu transformations (1.4) simplify to the Møller transformations by a change of time variable, i.e., $w = (1/\alpha_o)tanh(\alpha_o w^*)$.[4] The Møller transformations reduce to the identity transformations rather than the Lorentz transforma-tions in the limit of zero acceleration.

Experimentally, measurements of cosmic redshifts from faraway galaxies can be compared with calculations of the Wu-Doppler effect based on (1.4), treating the source as being at rest in an accelerated frame. The calculations are consistent with the observations.[6]

The physical space-time of a CLA frame is characterized by $P_{\mu\nu}$, which we call the Poincaré metric tensor,

$$P_{\mu\nu} = (W^2, -1, -1, -1), \qquad (1.5)$$

where $W(w, x)$ is given above. Analogous to how the Wu transformations reduce to the Poincaré transformations in the limit of zero acceleration, $P_{\mu\nu}$ reduces to the Minkowski metric tensor $\eta_{\mu\nu} = (1, -1, -1, -1)$, in the limit $\alpha_o \to 0$. The propagation of light in a CLA frame F is described by

$$ds^2 = W^2 dw^2 - dr^2 = 0, \qquad dr/dw = \pm W \neq constant, \qquad (1.6)$$

where $dr^2 = dx^2 + dy^2 + dz^2$. Thus, the speed of light in the CLA frame F is W, which is neither universal nor a constant.

We now face an urgent and non-trivial question: Do the space-time coordinates in CLA frames given by (1.4) have an operational meaning? Clearly, one can no longer synchronize clocks in the CLA frame, F_{CLA},

---

[b]In (1.4), we use the approximation $\gamma \approx \gamma_o(1+\beta_o\alpha_o w\gamma_o^2)$, which is obtained by expanding γ to the first order in α_o (which is assumed to be small), while keeping $\beta_o < 1$ to all orders.

using light signals as in inertial frames. Moreover, the spatial coordinates in (1.4) are no longer rigid, in contrast to those in inertial frames; they involve a distortion in the transformation. However, one could imagine constructing 'space-time clocks' for CLA frames and synchronizing them using the inverse transformations to (1.4). (See Eq. (3.7) in Ref. 2, Ch. 3.) A "space-time clock" is simply a device that shows a time and position and if based on, say, a computer chip, it could run at any specified rate and show any computable time and position. Thus, in order to synchronize clocks in the CLA frame, the space-time clocks associated with that frame would simply measure their own positions $\mathbf{r}_I = (x_I, y_I, z_I)$ relative to the F_I frame, obtain w_I from the nearest F_I clock, and then compute (using the known velocity β_o and acceleration α_o) and display w and $\mathbf{r} = (x, y, z)$ using the inverse of (1.4)[6] on a readout. Thus, the operational nature of space-time coordinates in non-inertial frames can be maintained as long as there exist explicit space-time coordinate transformations between non-inertial frames and inertial frames.

Some notable space-time properties of the CLA frame are as follows:

(a) The concept of constant linear acceleration implied by the Wu transformations is a constant change of a particle's energy per unit length, as measured by observers in an inertial frame. This is precisely the kind of acceleration experienced by a charged particle in a high energy linear accelerator.

(b) There is a singularity wall at $x_s = -1/(\alpha_o \gamma_o^2)$. The portion of space on one side of the singularity wall (i.e., $x < -1/(\alpha_o \gamma_o^2)$ has no correspondence to any portion of space in an inertial frame and, hence, is unphysical. In other words, observers in an inertial frame cannot detect the existence of anything on the other side of the singularity wall.

(c) This singularity wall retreats toward $-\infty$ as the acceleration α_o approaches zero.

(d) Clocks slow down $((dw/dw_I)_{dx=0} \to 0)$ and light travels faster $(\beta_c \to \infty)$ as one approaches the singularity wall.

(e) The time w is restricted by $(-1-\beta_o)/\alpha_o < w < (1-\beta_o)/\alpha_o$. Outside this range, the time w is unphysical and does not have operational meaning because it has no correspondence to any time w_I in an inertial frame.

(f) The set of transformations (1.4) form a pseudo-group[7] rather than the Lie group. It maps a portion of space of the CLA frame to the whole space of an inertial frame.

(g) The spatial coordinate of the CLA frame in (1.4) has a distortion and is not rigid, in contrast to that in an inertial frame. The generators of the CLA frames do not form a closed Lie algebra.

Having derived the Wu transformations (1.4), we now obtain the most general space-time coordinate transformations between an inertial frame $F_I(w_I, x_I, y_I, z_I)$ and a frame with an arbitrary linear acceleration $F_{ALA}(w, x, y, z)$. A minimal generalization of the Wu transformations is assumed to have the following transformations for the coordinate differentials,

$$dw_I = \gamma(W_a dw + \beta dx), \quad dx_I = \gamma(dx + \beta W_b dw), \quad dy_I = dy, \quad dz_I = dz;$$

$$\gamma = 1/\sqrt{1 - \beta^2}, \quad \beta = \beta_1(w) + \beta_o, \quad \beta^2 < 1, \tag{1.7}$$

$$W_a = \gamma^2 \alpha(w)x + \frac{\gamma^2}{\gamma_o^2} - \frac{\beta J_e}{\alpha^2(w)\gamma_o^2}, \quad W_b = \gamma^2 \alpha(w)x + \frac{\gamma^2}{\gamma_o^2} - \frac{J_e}{\beta \alpha^2(w)\gamma_o^2},$$

where $\alpha(w) = d\beta/dw$, $J_e = d\alpha(w)/dw$, and $W_a \approx W_b$ if the jerk J_e is small.

The principle of limiting continuation requires that the two functions W_a and W_b reduce to the same function W in the Wu transformations (1.4) in the limit of constant acceleration, i.e., as the arbitrary acceleration $\alpha(w)$ approaches a constant acceleration α_o, in addition to satisfying two integrability conditions.[4] For an arbitrary linear acceleration $\alpha(w)$, the resulting transformations are:

$$w_I = \gamma\beta\left(x + \frac{1}{\alpha(w)\gamma_o^2}\right) - \frac{\beta_o}{\alpha_o\gamma_o} + w_o,$$

$$x_I = \gamma\left(x + \frac{1}{\alpha(w)\gamma_o^2}\right) - \frac{1}{\alpha_o\gamma_o} + x_o,$$

$$y_I = y + y_o, \quad z_I = z + z_o; \tag{1.8}$$

$$\gamma = \frac{1}{\sqrt{1 - \beta^2}}, \quad \gamma_o = \frac{1}{\sqrt{1 - \beta_o^2}}, \quad \beta = \beta_1(w) + \beta_o,$$

where the acceleration $\alpha(w)$ and the velocity of the non-inertial frame $\beta(w)$ are both along parallel x and x_I axes.

A constant space-time shift x_o^μ has been included so that both the Lorentz and Poincaré transformations are special limiting cases of (1.8). We call (1.8) the "taiji space-time transformations" for a frame F_{ALA} with an acceleration $\alpha(w)$ that is an arbitrary function of w. We use the word "taiji" because in ancient Chinese philosophy, the word taiji describes the ultimate principle or condition that existed before the creation of the world. Here, "taiji" describes transformations between coordinates that are applicable before any constraints regarding accelerations or the universality of the speed of light are applied.

The inverse of the taiji transformation (1.8) is

$$\beta(w) = \frac{w_I + \beta_o/(\alpha_o\gamma_o) - w_o}{x_I + 1/(\alpha_o\gamma_o) - x_o} \equiv \beta_I(w_I, x_I),$$

$$x + \frac{1}{\alpha(w)\gamma_o^2} = \sqrt{\left(x_I + \frac{1}{\alpha_o\gamma_o} - x_o\right)^2 - \left(w_I + \frac{\beta_o}{\alpha_o\gamma_o} - w_o\right)^2}, \quad (1.9)$$

$$y = y_I - y_o, \quad z = z_I - z_o.$$

The arbitrary velocity function $\beta(w) = \beta_I(w_I, x_I)$ can always be expressed in terms of w_I and x_I, as shown in (1.9). However, in the most general case, the arbitrary acceleration function $\alpha(w)$ on the left-hand-side of (1.9) can not. Thus, the inverse taiji space-time transformations (1.9) cannot necessarily be written explicitly in the usual form $x^\mu = x^\mu(w_I, x_I)$. This is probably related to the fact that the ALA space-time transformations involving an arbitrary acceleration $\alpha(w)$, like the Wu transformations, form a pseudo-group rather than the usual Lie group because they map only a portion of space-time in the non-inertial frame to the whole of space-time in the inertial frame.[7] The generators of these pseudo-groups do not form a closed Lie algebra. However, it appears that Noether's theorem can be applied to physical theories in non-inertial frames because the number of generators in the transformations are countable infinite. This is in sharp contrast to those in general relativity, where the general coordinate transformations involve a continuously infinite number of generators. In her 1918 paper, Noether discussed a Theorem II, which showed that general relativity does not include the usual law for the conservation of energy.

In general, if a physical theory can be formulated in inertial frames, then it can be formulated in accelerated frames, whose metric tensors are known. In other words, if an equation is known in an inertial frame, the corresponding equation in an accelerated frame is obtained simply by replacing ordinary derivatives by covariant derivatives.[8] This rule can be applied to a higher derivative, because all inertial and non-inertial frames are in flat space-time.

1-4. Yang-Mills gravity

Yang-Mills gravity is based on the postulate that the 'external' space-time translational group is a dynamical group. This translation group does not have dimensionless and constant matrix representations and hence, it is not associated with a dimensionless coupling constant, in contrast to conventional internal gauge groups. However, these properties are actually just right for describing gravitational interactions because the gravitational (or T_4 gauge) and electromagnetic covariant derivative is given by

$$D_\mu = (\partial_\mu - ig\phi_{\mu\nu}p^\nu - ieA_\mu) = (\partial_\mu + g\phi_{\mu\nu}\partial^\nu - ieA_\mu). \tag{1.10}$$

This gauge covariant derivative, which appears in the Lagrangian of the electron (and positron) in the presence of quantum Yang-Mills gravity, has four important qualitative properties for all fields and quantum particles:

(i) The T_4 gauge covariant derivative (1.10) implies the universal gravitational coupling to all non-gravitational quantum fields. It also implies that all gauge symmetries and all renormalizable field theories are upset by the presence of quantum Yang-Mills gravity.

(ii) The T_4 gauge symmetry is the only exact symmetry in the physical world — unaffected by any other known interactions.

(iii) The gauge gravitational coupling constant g in (1.15) has the dimension of length and is naturally related to the Newtonian universal gravitational constant $G = g^2/(8\pi)$, and

(iv) The gravitational covariant derivative does not involve i, so that the gravitational force is consistently attractive for both particles and antiparticles, in contrast to the electromagnetic covariant derivative $\partial_\mu - ieA_\mu$, which involves an i.

Let us elaborate a bit more on the latter. We expect that a satisfactory theory of gravity should be able to explain why the gravitational force is only attractive and never repulsive.[2,6] This property can be seen in the coupling between the gravitational tensor field $\phi_{\mu\nu}$ and the matter fermion field ψ at the quantum level in the Lagrangian. Let us consider the gravitational (T_4) tensor field $\phi_{\mu\nu}(x)$ and the electromagnetic potential field $A_\mu(x)$ in the gauge covariant derivative and its complex conjugate in the gravitational Dirac equation,[2]

$$\partial_\mu - ig\phi_\mu^\nu p_\nu - ieA_\mu + ... = \partial_\mu + g\phi_\mu^\nu \partial_\nu - ieA_\mu + ... \tag{1.11}$$

$$(\partial_\mu - ig\phi_\mu^\nu p_\nu - ieA_\mu + ...)^* = \partial_\mu + g\phi_\mu^\nu \partial_\nu + ieA_\mu + ... \tag{1.12}$$

The gauge covariant derivative (1.11) and its complex conjugate (1.12) appear respectively in the wave equations of the electron (i.e., a particle with charge $e < 0$) and the positron (i.e., an antiparticle with charge $-e > 0$). The key properties of the electric force $F_e(e^-, e^-)$ (i.e., between electron and electron) and the force $F_e(e^-, e^+)$ (i.e., between electron and positron) are given by the third terms in (1.11) and in (1.12), i.e.,

$$F_e(e^-, e^-): \quad (-ie) \times (-ie) = -e^2, \quad \textit{repulsive}, \tag{1.13}$$

$$F_e(e^-, e^+): \quad (-ie) \times (ie) = +e^2, \qquad attractive, \qquad (1.14)$$

and the force $F_e(e^+, e^+)$ is the same as $F_e(e^-, e^-)$. Thus, we have experimentally established attractive and repulsive electric forces, which are due to the presence of i in the electromagnetic coupling. The Yang-Mills gravitational force $F_{YMg}(e^-, e^-)$ (i.e., between electron and electron) and the force $F_{YMg}(e^-, e^+)$ (i.e., between electron and positron) are given by the second term in (1.11) and in (1.12), respectively. Because the gravitation coupling terms in (1.11) and (1.12) do not involve i, we have only an attractive gravitational force,

$$F_{YMg}(e^-, e^-): \quad (g) \times (g) = +g^2, \qquad attractive, \qquad (1.15)$$

$$F_{YMg}(e^-, e^+): \quad (g) \times (g) = +g^2, \qquad attractive, \qquad (1.16)$$

and $F_{YMg}(e^+, e^+)$ is the same as $F_{YMg}(e^-, e^-)$. Note that these qualitative results for the forces $F_e(e^-, e^-)$ and $F_{YMg}(e^-, e^-)$ are independent of the signs of the coupling constants e and g. This is important because the intrinsic nature of force should not depend on our convention for the signs of coupling constants. These considerations hold only at the level of quantum gravity and can be generalized to other quantum particles and antiparticles. Classical Yang-Mills gravity is based on the macroscopic Hamilton-Jacobi equations, $G^{\mu\nu}(\partial_\mu S)(\partial_\nu S) - m^2 = 0$, or geodesic equations, which cannot provide an understanding of the attractive nature of the gravitational force for all particles and antiparticles.

Furthermore, the generators of the space-time translation group T_4 are $p_\mu = i\partial_\mu$, which have the dimension of $(1/\text{length})$ in natural units. Therefore, the gravitational coupling constant g in the T_4 gauge covariant derivatives (1.11) and (1.12) has the dimension of length (in natural units), in sharp contrast to all other coupling constants of fields associated with internal gauge groups, so that g^2 is related to the Newtonian constant G by $g^2 = 8\pi G$. These important qualitative results reveal that the space-time translation T_4 gauge group of Yang-Mills gravity is just right for gravity.

1-5. General Yang-Mills symmetry with vector gauge functions and Hamilton's characteristic phase factors

E. P. Wigner sorted symmetries in physics into two classes: *Geometric Symmetries*, which lead to conservation laws, and *Dynamic Symmetries*, which dictate the dynamics of interactions in physics, in addition to having a corresponding conservation law.[3] One of the simplest dynamic symmetries is the U_1 gauge symmetry associated with the electric charge. The U_1

gauge transformations of the electromagnetic field $A_\mu(x)$ and the election field $\psi(x)$ are as follows,

$$A'_\mu(x) = A_\mu(x) + \partial_\mu\omega(x), \qquad \psi'(x) = e^{-ie\omega(x)}\psi(x), \qquad (1.17)$$

where $\omega(x)$ is an arbitrary scalar gauge function. Analogous to the invariance of the spacetime interval under the Lorentz transformations, the electromagnetic field strength $F_{\mu\nu}(x) = \partial_\mu A_\nu(x) - \partial_\nu A_\mu(x)$ is invariant under the U_1 gauge transformation, i.e.,

$$F'_{\mu\nu} = \partial_\mu A'_\nu(x) - \partial_\nu A'_\mu(x) = F_{\mu\nu}. \qquad (1.18)$$

Thus, the electromagnetic action in its usual quadratic form,

$$S_{em} = -\frac{1}{4}\int F_{\mu\nu}(x)F^{\mu\nu}(x)d^4x, \qquad (1.19)$$

is also invariant under the U_1 gauge transformation (1.10). Maxwell's equations can then be obtained from the gauge invariant action S_{em} and a gauge invariant identity of $F_{\mu\nu}$,

$$\partial_\lambda F_{\mu\nu} + \partial_\mu F_{\nu\lambda} + \partial_\nu F_{\lambda\mu} = 0. \qquad (1.20)$$

It now appears clear that dynamical symmetries of the physical world are gauge symmetries. They have far-reaching consequences and their importance to foundational physics cannot be emphasized strongly enough.

Similar to the problem with spacetime frameworks in which the curved spacetime of Einstein's general relativity is incompatible with the flat spacetime for understanding quantum mechanical phenomena, quantum chromodynamics (QCD) has a basic difficulty in producing an explicit mechanism for confining quarks. Although conventional QCD has an SU_3 gauge symmetry associated with color charges, it does not give rise to a linear potential in the static limit that is compatible with the observed charmonium spectrum[9] and the permanent confinement of quarks and gluons. This is in sharp contrast to quantum electrodynamics (QED) that does give rise to the Coulomb potential in the static limit, which is compatible with measurements of atomic energy spectra.

One natural way to obtain a linear potential to realize quark confinement explicitly is to generalize the usual gauge (or Yang-Mills) symmetries to something we call 'general Yang-Mills symmetries.' Analogous to our approach of making a minimum generalization to the Poincaré transformations in order to obtain coordinate transformations that can be applied to accelerated frames, our approach here is to replace the scalar gauge function $\omega(x)$ in the gauge transformation by a (Lorentz) vector function $\omega_\mu(x)$.

Analogous to the case where the transformations for accelerated frames reduce smoothly to the Poincaré transformations in the limit of zero acceleration, general Yang-Mills symmetries reduce to the conventional Yang-Mills symmetries in the special case where the (Lorentz) vector gauge function $\omega_\mu^a(x)$ can be expressed as the space-time derivative of a (Lorentz) scalar gauge function $\omega^a(x)$,

$$\omega_\mu^a(x) = \partial^\mu \omega^a(x). \tag{1.21}$$

This property (1.21) is important because it enables us to treat the unified electroweak theory with the usual $SU_2 \times U_1$ symmetries involving spontaneous symmetry breaking as a special case of the general Yang-Mills symmetry. This relation for general Yang-Mills symmetry and usual gauge symmetry holds for both Abelian and non-Abelian groups.

This more general type of symmetry leads to new fourth-order field equations, which can be used to obtain dual linear and Coulomb-like potentials in the static limit. Such dual potentials are the same as the empirical dual potentials obtained from the analysis of the data of the spectrum of charmonium.[9]

A second advantage of this approach is that we can also generalize the original Lee-Yang U_1 symmetry to baryon charges. The general U_1 transformations corresponding to (1.17) for a baryonic gauge field $B_\mu(x)$ and fermion field $\Psi(x)$, which carries a baryon charge g_b, are assumed to be

$$B_\mu'(x) = B_\mu(x) + \Lambda_\mu(x), \qquad \Psi(x) = \Omega \Psi(x), \tag{1.22}$$

$$\Omega = exp\left[-ig_b \int_{x_o'}^{x_e'^\lambda = x} dx'^\lambda \Lambda_\lambda(x')\right]_{Le_B},$$

where Ω is Hamilton's characteristic phase factor[10,11] and

$$\partial^\mu B_{\mu\nu}'(x) = \partial^\mu B_{\mu\nu}(x), \qquad B_{\mu\nu}(x) = \partial_\mu B_\nu(x) - \partial_\nu B_\mu(x), \tag{1.23}$$

provided the following constraint is imposed,

$$\partial^\mu \partial_\mu \Lambda_\nu(x) - \partial^\mu \partial_\nu \Lambda_\mu(x) = 0, \tag{1.24}$$

where $\Lambda_\mu(x) \neq \partial_\mu \xi(x)$ and Λ_μ are vector gauge functions and $\xi(x)$ is a usual scalar function. The solutions of the second-order partial differential equations in (1.24) form a set of infinitely many functions.

The subscript 'Le_B' in (1.22) denotes that the path in the integration is required to satisfy the equation,

$$[\partial_\mu \Lambda_\lambda(x) - \partial_\lambda \Lambda_\mu(x)]dx^\mu = 0, \tag{1.25}$$

so that the phase factor Ω is a local function of x, and we have

$$\partial_\mu \Omega = -ig_b \Lambda_\mu(x)\Omega. \tag{1.26}$$

This equation corresponds to the relation $\partial_\mu exp(-ie\omega) = -ie(\partial_\mu \omega) \times exp(-ie\omega)$ in the usual U_1 gauge symmetry. It is necessary for general U_1 symmetry of a Lagrangian involving $\Psi(x)$ in (1.22).

Thus, the new general Yang-Mills transformations involve (Lorentz) vector gauge functions (as opposed to scalar gauge functions) and Hamilton's characteristic phase factors.

In summary, a key new underpinning of Broader Particle-Cosmology is general Yang-Mills (YM) symmetry, which leads to fourth-order gauge field equations and consequently, static linear and Coulomb-like potentials. As we shall see in later chapters, such potentials for color charges provide an explicit mechanism for confining quarks and for understanding the charmonium spectra. Moreover, such potentials for baryonic charges imply the existence of a linear force on the cosmic scale that can provide a field-theoretic understanding of the late-time acceleration of cosmic expansion. Thus, Broader Particle-Cosmology in flat space-time accommodates all possible fundamental forces in nature, i.e., the strong force, the electroweak force, and the gravitational force, as well as the not-yet-observed baryon and lepton forces.

References

1. See, for examples, P. A. Zyla et al. (Particle Data Group), Prog. Theor. Exp. Phys. 2020, 083C01 (2020); T. D. Lee, *Particle Physics and Introduction to Field Theory* (Hardwood Academic, 1981) pp. 746–752; K. Huang, *Quarks, Leptons and Gauge Fields* (World Scientific, 2nd ed., 1992). H. Y. Chiu, Phys. Rev. Letter. 17, 712, (1966); Ya. B. Zel'dovich, Advances Astron. Astrophys. 3 242 (1965).
2. J. P. Hsu and L. Hsu, *Space-Time, Yang-Mills Gravity, and Dynamics of Cosmic Expansion* (World Scientific, 2020), Ch. 3, p. 16, p. 80; Phys. Lett. A **196** 1 (1994); L. Hsu and J. P. Hsu, Nuovo Cimento B **111**,1283 (1996).
3. F. J. Dyson, Bulletin of the American Math. Soc., 78, Sept. 1972. See also J. P. Hsu and D. Fine, *100 Years of Gravity and Accelerated Frames, The Deepest Insight of Einstein and Yang-Mills* (World Scientific, 2005) pp. 347–352, (with A Brief Remark for 'Missed Opportunity' by Dyson). Dyson stressed: "The most glaring incompatibility of concepts in contemporary physics is that between Einstein's general coordinate invariance and all the modern schemes for a quantum mechanical description of nature". E. P. Wigner, *Symmetries and Reflections, Scientific Essays* (The MIT Press, 1967), pp. 52–53 (in collaboration with H. Salecker). Wigner wrote "Evidently, the usual statements about future positions of particles, as specified

by their coordinates, are not meaningful statements in general relativity. This is a point which cannot be emphasized strongly enough and is the basis of a much deeper dilemma than the more technical question of the Lorentz invariance of the quantum field equations. It pervades all the general theory, and to some degree we mislead both our students and ourselves when we calculate, for instance, the mercury [sic] perihelion motion without explaining how our coordinate system is fixed in space.... Expressing our results in terms of the values of coordinates became a habit with us to such a degree that we adhere to this habit also in general relativity, where values of coordinates are not per se meaningful."

4. C. Møller, Danske Vid. Sel. Mat-Fys. **20**, No. 19 (1943); and *The Theory of Relativity* (Oxford Univ. Press, London, 1952), pp. 253–258; H. Lass, Am. J. Phys. **31**, 274 (1963); W. Rindler, Am. J. Phys. **34**, 1174 (1966); R. A. Nelson, J. Math. Phys. **28**, 2379 (1987); S. G. Turyshev, O. L. Minazzoli, and V. T. Toth, J. Math. Phys. **53**, 032501 (2012); T.-Y. Wu and Y. C. Lee, Int. J. Theor. Phys. **5**, 307 (1972).

5. J. P. Hsu and L. Hsu, Eur. Phys. J. Plus (2013) **128**: 74; and *Space-Time, Yang-Mills Gravity and Dynamics of Cosmic Expansion* (World Scientific, 2013), Ch. 2–3, pp. 36–38.

6. L. Hsu and J. P. Hsu, Chinese Phys. C. **43** 105103 (2019); Ref. 2, pp. 26–27 and pp. 39–43; J. P. Hsu, Chinese Phys. C. **41** 015101 (2017); J. P. Hsu, arXiv:1911.10901; S. Weinberg, *Cosmology* (Oxford Univ. Press, 2008) p. 39.

7. O. Veblen and J. H. C. Whitehead, *The Foundations of Differential Geometry* (Cambridge Univ. Press, 1953), pp. 37–38. (It appears that the idea of pseudo-group was also discussed earlier by E. Cartan.) Physically, the pseudo-group transformations map a portion of space-time in one non-inertial frame to a portion of space-time in another non-inertial frames. Moreover, the generators of a pseudo group do not form a closed Lie algebra, in contrast to those of Lorentz groups. (This property can be verified by explicit calculations.) E. Noether, Goett. Nachr., 235 (1918). English translation of Noether's paper is online. Google search: M. A. Tavel, Noether's paper.

8. V. Fock, *The Theory of Space Time and Gravitation* (tr. by N. Kemmer, Pergamon Press, 1958) p. 149.

9. E. Eichten, K. Gottfried, T. Kinoshita, J. Kogut, K. D. Lane, and T.-M. Yan, Phys. Rev. Lett. 34, 369 (1975).

10. W. Yourgrau and S. Mandelstam, *Variational Principle in Dynamics and Quantum Theory* (Dover, 3rd edition, 1979) p. 50; L. Landau and E. Lifshitz, *The Classical Theory of Fields* (Addison-Wesley, 1951) p. 29.

11. L. Hsu and J. P. Hsu, Chin. Phys. C, Vol. 43, No. 10, 105103 (2019). See Ch. 6, Sec. 6-2.

2

A Model for Dark Matter

2-1. Introduction

According to the conventional view, it appears that much of the mass in the universe is composed of unobserved non-baryonic matter that interacts only gravitationally with luminous (visible) matter. By some estimates, the amount of dark matter in the universe is 5 to 60 times larger than the amount of ordinary matter. It is no wonder that a central problem in current physics is to determine the constituents of this 'dark' matter. Some ad hoc ideas that have been proposed involve as-yet-unobserved particles (such as a very heavy flavor of neutrino) and/or new laws of physics.[1]

However, in the spirit of hewing closely to known particles and the established laws and gauge symmetries of modern particle physics, we propose a high energy neutrino (HEN) model for dark matter, based on Broader Particle-Cosmology.

We propose that the observed equal numbers of protons and electrons in our universe implies that there should also be an equal number of anti-electron-neutrinos (anti-e-neutrinos $\bar{\nu}_e$). The gravitational interaction of these proposed anti-e-neutrinos with luminous matter is enhanced due to the relativistic properties of the energy-momentum tensor in Yang-Mills gravity[1] (a similar effect is present in general relativity). As a result, the effects of their gravitational interaction with luminous matter is much larger than the interaction strength one might expect from their rest mass of roughly 1 eV. In addition, gauge symmetry and the observed conservation of lepton charge suggests the existence of a lepton-lepton force which, though almost too small to observe on ordinary scales, could have significant effects on a galactic level. We propose the combination of these effects as the cause for phenomena such as the observed flat rotation curves of galaxies.

Based on known conservation laws and gauge symmetries, the HEN model for dark matter is based on three ideas:

1. There are many more anti-electron-neutrinos (or anti-e-neutrino, $\bar{\nu}_e$) in the universe than is conventionally assumed. However, the much larger number of anti-e-neutrinos that we propose is still consistent with established physics and is a consequence of a slightly different assumption about conditions near the beginning of the universe.

2. Yang-Mills gravity implies that, for high energy neutrinos and anti-neutrinos, the gravitational interaction is enhanced by a factor equal to the ratio of the energy of the neutrino to the rest mass of the neutrino, which could be larger than 10^6. Since Yang-Mills gravity is consistent with many of the predictions of general relativity, it is not surprising that general relativity predicts a similar phenomenon.

3. There is an as-yet-unobserved lepton-lepton force that, paired with the excess of anti-e-neutrinos, leads to the effects currently attributed to dark matter. This lepton-lepton force is not new physics, but a natural extension of gauge symmetry applied to the conservation of lepton charge.

2-2.   The creation of anti-e-neutrinos in the early universe

The conventional picture of the beginning of the universe is that conditions were similar to that inside a gigantic super-collider. The typical strong interaction time τ_{int} between two quarks is about the same or shorter than that between two hadrons in a high energy collision event, i.e., roughly $(1 \ fermi)/(speed \ of \ light) \approx 10^{-14}s \approx \tau_{int}$, so the time interval between the creation of quarks and antiquarks and the appearance of hadrons is approximately on the order of this interaction time. Based on observations that the majority of the visible universe is composed of hydrogen, i.e., protons and electrons, and that the universe seems to be neutral, we make the natural assumption that the dominant process during quark confinement at the beginning of the universe was the formation of neutrons and N resonances.[a]

Since N resonances have an extremely short lifetime, the baryonic matter of the universe would thus quickly come to be dominated by neutrons, which would then decay via the weak interaction

$$n \rightarrow p^+ + e^- + \overline{\nu}_e,$$

with a comparatively long lifetime on the order of $10^3 s$. Consequently, roughly a few thousand seconds after the creation of the universe, we would expect most of these neutrons to have decayed into the protons and electrons that we see today, along with an equal number of anti-e-neutrinos. Of course, there are many other non-dominant decay processes that lead to comparatively tiny non-equal numbers of protons and electrons. However, of key importance is the result that anti-e-neutrinos should be as numerous as protons and electrons.

---

[a]We address the issue of the observed relative abundances of matter and anti-matter in Ch. 4.

Because the HEN model assumes that these anti-e-neutrinos were created near the beginning of the universe and at creation, we would expect the hadrons to have extremely high energies, it is natural to assume that the weak interaction would produce anti-e-neutrinos with extremely high energies as well. In Sec. 2-6, we discuss possible observational evidence involving cosmic ray positrons that suggests that these anti-e-neutrinos have high energies.

One difference between the HEN model and the standard Big Bang model is that due to the relatively long lifetime of the neutron, the time of synthesis of protons in the HEN model is estimated to be on the order of thousands of seconds after creation, much later than that in the standard big bang model. However, this HEN model remains consistent with the fact that there are almost equal numbers of electrons and protons in the observable cosmos.

2-3. Yang-Mills gravity and the enhanced gravitational interaction of high energy anti-e-neutrinos

Particle physics and quantum field theories are closely related to the Poincaré transformations (or the Poincaré group) in flat space-time. However, general relativity is incompatible with the Poincaré group. Thus, there is a conceptual difficulty for particle-cosmology if general relativity is used for cosmology. Dyson[7] stressed that 'The most glaring incompatibility of concepts in contemporary physics is that between Einstein's principle of general coordinate invariance and all the modern schemes for a quantum-mechanical description of nature.' However, quantum Yang-Mills gravity, being formulated on the basis of translation gauge symmetry in flat space-time, is compatible with the Poincaré group.[1,3,4]

Although the consistency of Yang-Mills gravity with experimental measurements has been discussed elsewhere,[1,7] here we simply point out that in the geometric-optics or classical limit, the quantum wave equation of fermions field in Yang-Mills gravity reduces to a Hamilton-Jacobi type equation, which we call the 'Einstein-Grossmann' equation of motion[1]

$$G^{\mu\nu}(\partial_\mu S)(\partial_\nu S) - m^2 = 0, \qquad G^{\mu\nu} = \eta_{\alpha\beta}J^{\alpha\mu}J^{\beta\nu}. \qquad (2.1)$$

This is the basic dynamics equation for classical Yang-Mills gravity. It dictates the motion of classical objects and light rays in flat space-time, and involves a new effective Riemannian metric tensor $G^{\mu\nu}$. This 'Einstein-Grossmann' equation of motion of Yang-Mills gravity is formally the same as the corresponding dynamics equation for macroscopic objects in general relativity.[7] Thus, the reason why Yang-Mills gravity and general relativity make similar experimental predictions is that all the phenomena that can be explained by a curvature of space-time can be explained by translational gauge symmetry for the motion of quantum particles in a flat space-time in the classical limit equally well. Equation (2.1) enables Yang-Mills gravity to make correct predictions concerning the perihelion shift of the Mercury, the deflection of light by the sun, and other phenomena attributed to curved space-time.[7]

To calculate the gravitational interaction between high energy anti-e-neutrinos and other matter, let us consider general time-independent (static) approximations for the gravitational potential in flat space-time based on Yang-Mills gravity. In inertial frames with the metric $\eta_{\mu\nu} = (1, -1, -1, -1)$, it suffices to consider the linearized gauge field wave equation,

$$\partial_\lambda \partial^\lambda \phi^{\mu\nu} - \partial^\mu \partial_\lambda \phi^{\lambda\nu} - \eta^{\mu\nu} \partial_\lambda \partial^\lambda \phi + \eta^{\mu\nu} \partial_\alpha \partial_\beta \phi^{\alpha\beta}$$

$$+\partial^\mu \partial^\nu \phi - \partial^\nu \partial_\lambda \phi^{\lambda\mu} - gS^{\mu\nu} = 0. \tag{2.2}$$

The source $gS^{\mu\nu}$ by itself cannot generate the (00) field component, ϕ^{00}. Nevertheless, this equation can be written in the following form:[4]

$$\partial_\lambda \partial^\lambda \phi^{\mu\nu} - \partial^\mu \partial_\lambda \phi^{\lambda\nu} + \partial^\mu \partial^\nu \phi_\lambda^\lambda - \partial^\nu \partial_\lambda \phi^{\lambda\mu} = g\left(S^{\mu\nu} - \frac{1}{2}\eta^{\mu\nu} S_\lambda^\lambda\right). \tag{2.3}$$

As with the Einstein-Grossman equation of motion, the linearized gauge-field Eq. (2.3) is mathematically the same as the corresponding linearized equation in general relativity. The ϕ^{00} component in (2.3) leads to the gravitational potential in Yang-Mills gravity.[1]

In the relativistic static approximation, we note that the source term $S^{\mu\nu}$ in (2.3) transforms in the same way as the energy-momentum tensor $T^{\mu\nu}$ of a classical particle. Thus, we express the source as follows:

$$g\left(S^{\mu\nu} - \frac{1}{2}\eta^{\mu\nu} S_\lambda^\lambda\right) \approx g\left(T^{\mu\nu} - \frac{1}{2}\eta^{\mu\nu} T_\lambda^\lambda\right). \tag{2.4}$$

In analogy to charge density, one can write either[5]

$$(A): \quad T^{\mu\nu} = \rho \frac{dx^\mu}{ds} \frac{dx^\nu}{dt}, \quad ds = \sqrt{1 - \beta^2} \, dt, \tag{2.5}$$

or

$$(B): \quad T^{\mu\nu} = \rho_o \frac{dx^\mu}{ds} \frac{dx^\nu}{ds}, \quad \rho_o = \rho\sqrt{1 - \beta^2},$$

where

$$T_\lambda^\lambda = \frac{\rho E}{m}(1 - \beta^2), \quad E/m = 1/\sqrt{1 - \beta^2}. \tag{2.6}$$

As can be seen from (2.4)–(2.6), it is possible to make two different high energy approximations for the source tensor $T^{\mu\nu}$ in the non-Lorentz invariant static limit:

$$(A): \ \rho \to m\delta^3(\mathbf{r}), \quad T^{00} = \rho \frac{dx^0}{ds} \frac{dx^0}{dt} \to m(E/m)\delta^3(\mathbf{r}), \tag{2.7}$$

$$(B): \ \rho_o \to m\delta^3(\mathbf{r}), \quad T^{00} = \rho_o \frac{dx^0}{ds} \frac{dx^0}{ds} \to m(E/m)^2\delta^3(\mathbf{r}), \tag{2.8}$$

where $t = x^0$. Note that both approximations for T^{00} in (2.7) and (2.8) reduce to the usual non-relativistic static limit $T^{00} \to m\delta(\mathbf{r})$ when $E \to m$. Thus far, there are no experiments that can distinguish between these two different approximations. Consequently, we shall explore the cosmic implications of both. Based on (2.3)–(2.8), we write

$$\nabla^2 \phi^{00} = -(1/2)gE(1 + \beta^2)\delta^3(\mathbf{r}), \quad for \ \ \rho \to m\delta^3(\mathbf{r}); \tag{2.9}$$

$$\nabla^2 \phi^{00} = -(1/2)gE(E/m)(1 + \beta^2)\delta^3(\mathbf{r}), \quad for \ \ \rho_o \to m\delta^3(\mathbf{r}).$$

In Yang-Mills gravity based on translational gauge symmetry in flat space-time, the motion of a classical particle is described by the Einstein-Grossman equation of motion[7]

$$G^{\mu\nu}(\partial_\mu S)(\partial_\nu S) - m^2 = 0, \quad G^{\mu\nu} = \eta_{\alpha\beta}(\eta^{\alpha\mu} + g\phi^{\alpha\mu})(\eta^{\beta\nu} + g\phi^{\beta\nu}). \tag{2.10}$$

In the classical limit, ϕ^{00} plays the role of the static Newtonian potential energy. Equations (2.9) lead to

$$\phi^{00} = (1 + \beta^2)\frac{gE}{8\pi r}, \quad for \ \ \rho \to m\delta^3(\mathbf{r}), \tag{2.11}$$

$$\phi^{00} = (1 + \beta^2)\frac{gE(E/m)}{8\pi r}, \quad for \ \ \rho_o \to m\delta^3(\mathbf{r}).$$

Thus, the Yang-Mills gravitational forces $(-d/dr)[-M'g\phi^{00}]$ between a relativistic neutrino (or anti-neutrino) with energy $E = E_\nu$ (mass $m = m_\nu$) and a macroscopic body with mass M' are given by

$$(A): \ \ F_{YM} = \frac{-GM'E_\nu(1 + \beta^2)}{r^2}, \quad for \ \ \rho \to m_\nu\delta^3(\mathbf{r}); \tag{2.12}$$

$$(B): \ \ F_{YM} = \frac{-GM'E_\nu(E_\nu/m_\nu)(1 + \beta^2)}{r^2}, \quad for \ \ \rho_o \to m_\nu\delta^3(\mathbf{r});$$

where m_ν is the neutrino mass, and G is the Newtonian constant, $G = g^2/8\pi$.

Thus, although the usual gravitational force between a neutrino at rest and another object is negligible because the neutrino mass is extremely small, i.e., $m_\nu < 1.1\ eV$,[8] the static gravitational force between a relativistic neutrino with $E_\nu > 1MeV$ and another object may no longer be negligible because E_ν/m_ν could be larger than 10^6.

As a point of comparison, general relativity provides for a similar enhancement. The force between a relativistic electron (or a photon) with energy E and velocity $\boldsymbol{\beta} = \mathbf{v}/c$ and a macroscopic body with mass M is

$$\mathbf{F}_g = \frac{-GME[\mathbf{r}(1 + \beta^2) - \boldsymbol{\beta}(\boldsymbol{\beta} \cdot \mathbf{r})]}{r^3}, \qquad c = \hbar = 1, \qquad (2.13)$$

as derived by Okun.[6]

2-4.   The leptonic Lee-Yang force

In 1955, Lee and Yang discussed a local U_1 gauge symmetry associated with the conservation of baryon charge (or number).[3, 10] They showed that the dynamic equations of the baryonic U_1 gauge field are formally the same as Maxwell's equations. Based on *Eötvös's* experiment,[10] Lee and Yang estimated the new baryonic force to be weaker than the gravitational force by a factor of 10^{-5} or smaller.

However, Lee and Yang's idea can also be applied to the conserved leptonic charge, implying that there should be an inverse-square force between leptons, in addition to the long range Coulomb and gravitational forces and the short range weak force. Because such a leptonic Lee-Yang force and its U_1 gauge field have not been detected in high energy experiments, some physicists consider the U_1 symmetry corresponding to leptonic charge not to be a local gauge symmetry. However, particle physics strongly suggests the primacy of the gauge symmetry principle in the physical world,[3, 4, 7] which is supported by the success of QED, QCD and the unified electroweak theory. Thus, it is reasonable for us to postulate the principle of universal gauge symmetry:

All conserved charges are associated with local internal gauge symmetries.[b]

Although this as-yet-undetected leptonic force is undoubtedly very weak, it could play a role in cosmic phenomena where an extremely large number of leptons is involved. In particular, the inverse-square dependence

---

[b]To include gravity, the local external space-time translation gauge symmetry for quantum Yang-Mills gravity[1] should be incorporated. These gauge symmetries are dynamic symmetries.

of this force gives it the right properties for supplementing the gravitational force and, in combination with high energy anti-e-neutrinos that are as numerous as protons and electrons, for producing the observed flat-rotation curves of galaxies.

Although Yang-Mills gravitational interaction of high-energy neutrinos has been discussed in the previous section, in the following derivation, we write the relevant Lagrangian with both leptonic gauge symmetry and space-time translational gauge symmetry to demonstrate the ability of the framework of Broader Particle-Cosmology to incorporate both of these apparently unrelated forces, similar to the unified electroweak theory.

In an inertial frame, the Lagrangian $L_{\ell\phi}$ involving a set of leptons $\ell = (e, \nu_e)$, the leptonic U_1 gauge field $L_\mu(x)$, and gravitational spin-2 fields[4] $\phi^{\mu\nu}(x)$ takes the form,

$$S_{\ell\phi} = \int L_{\ell\phi} d^4 x, \qquad c = \hbar = 1, \tag{2.14}$$

$$L_{\ell\phi} = i[\bar{\ell}\gamma_\mu \Delta^\mu \ell] - m_\ell \bar{\ell}\ell - \frac{1}{4} L_{\mu\nu} L^{\mu\nu},$$

$$+ \frac{1}{2g^2} \left(C_{\mu\alpha\beta} C^{\mu\beta\alpha} - C_{\mu\alpha}{}^\alpha C^{\mu\beta}{}_\beta \right), \qquad g^2 = 8\pi G, \tag{2.15}$$

$$C^{\mu\nu\lambda} = J^{\mu\sigma} \partial_\sigma J^{\nu\lambda} - J^{\nu\sigma} \partial_\sigma J^{\mu\lambda}, \tag{2.16}$$

$$J^{\mu\nu} = \eta^{\mu\nu} + g\phi^{\mu\nu} = J^{\nu\mu}, \tag{2.17}$$

$$\Delta^\mu \ell = (\partial^\mu + g\phi^{\mu\nu}\partial_\nu - ig_\ell L^\mu)\ell, \qquad L_{\mu\nu} = \partial_\mu L_\nu - \partial_\nu L_\mu, \tag{2.18}$$

$$C_{\mu\alpha\beta} C^{\mu\beta\alpha} = (1/2) C_{\mu\alpha\beta} C^{\mu\alpha\beta} \tag{2.19}$$

where G is the universal gravitational constant and $\eta^{\mu\nu} = (1, -1, -1, -1)$. The gauge-fixing terms[4] for $\phi^{\mu\nu}$ and L_μ are not included because we do not consider quantization of fields here.

From the gauge invariant action (2.14), one can derive equations for the leptonic gauge fields $L_\mu(x)$ and the gravitational tensor field $\phi^{\mu\nu}(x)$,

$$\partial^\mu L_{\mu\nu} + g_\ell \bar{\ell}\gamma_\nu \ell = 0, \tag{2.20}$$

$$H^{\mu\nu} = g^2 S^{\mu\nu}, \tag{2.21}$$

$$H^{\mu\nu} = \partial_\lambda (J^\lambda_\rho C^{\rho\mu\nu} - J^\lambda_\alpha C^{\alpha\beta}{}_\beta \eta^{\mu\nu} + C^{\mu\beta}{}_\beta J^{\nu\lambda})$$

$$- C^{\mu\alpha\beta} \partial^\nu J_{\alpha\beta} + C^{\mu\beta}{}_\beta \partial^\nu J^\alpha_\alpha - C^{\lambda\beta}{}_\beta \partial^\nu J^\mu_\lambda, \tag{2.22}$$

where the symmetric[4] 'source tensor' $S^{\mu\nu}$ for fermion matter is given by

$$S^{\mu\nu} = \bar{\ell} i\gamma^{\mu}\partial^{\nu}\ell. \tag{2.23}$$

The gauge fields L_{μ} are generated by leptonic charges. The wave equation of a lepton is given by

$$[i\gamma_{\mu}(\partial^{\mu} + g\phi^{\mu\nu}\partial_{\nu} - ig_{\ell}L^{\mu}) - m_{\ell}]\,\ell = 0. \tag{2.24}$$

Based on this quantum equation and following the steps (1.11)–(1.16) in Ch. 1, we can demonstrate that the leptonic force has both attractive and repulsive properties; while the Yang-Mills gravitational force is always attractive for all leptons and anti-leptons.

As has been discussed elsewhere,[1,7,9] we assume the external spacetime translation gauge group T_4 for gravity. Its gauge covariant derivative is very similar to that of electrodynamics with U_1 gauge symmetry. The universal gravitational coupling to all physical fields in Yang-Mills gravity is realized through the replacement of the ordinary partial derivative ∂_{μ} by the T_4 gauge covariant derivative $(\partial^{\mu} + g\phi^{\mu\nu}\partial_{\nu})$ in the Lagrangians of all non-gravitational fields.

Let us consider the relativistic static potential for the leptonic Lee-Yang force. In the static limit, where $L_{\mu} = L_{\mu}(r)$, Eq. (2.20) with $\mu = 0$ can be written as

$$[\partial^2 L_{\mu} - \partial_{\mu}\partial^{\nu}L_{\nu}]_{\mu=o} = -\nabla^2 L_0(r) = -g_{\ell}\bar{\ell}\gamma_0\ell, \tag{2.25}$$

where the relativistic neutrino source term in (2.25), i.e., $\ell = \nu$, transforms like the zeroth component of a 4-vector, so that we have the static approximation, $\bar{\ell}\gamma_0\ell \to (E_{\nu}/m_{\nu})\delta^3(\mathbf{r})$.[1,2]

The zeroth component static gauge potential $L_{\mu}(r), \mu = 0$, then satisfies the equation in such a 'relativistic static limit',

$$\nabla^2 L_0(r) \approx g_{\ell}(E_{\nu}/m_{\nu})\delta^3(\mathbf{r}), \tag{2.26}$$

which is consistent with (2.12) and (2.23) for a static force or potential produced by a relativistic particle. The relativistic static Eq. (2.26) leads to a Coulomb-like potential energy $g_{\ell}L_0(\mathbf{r}) = -g_{\ell}^2(E_{\nu}/m_{\nu})/4\pi r$. Thus, the attractive Lee-Yang force F_{LY} between a non-relativistic electron (i.e., $E_e/m_e \approx 1$) and a relativistic anti-e-neutrino with opposite leptonic charge and with large energy $(E_{\nu}/m_{\nu}) >> 1$ is given by

$$F^{LY} = -g_{\ell}\frac{dL_0}{dr} = \frac{-g_{\ell}^2(E_{\nu}/m_{\nu})}{4\pi r^2}. \tag{2.27}$$

2-5. Flat rotation curves due to enhanced gravity and the cosmic Lee-Yang force on high energy neutrinos

We now discuss how the HEN model can produce a flat rotation curve for galaxies. Although the Lee-Yang force produces a repulsive interaction between all the electrons in a galaxy and between all the anti-e-neutrinos, we neglect those effects in the following discussion because: (1) the repulsive force between the electrons will be negligible compared to the electron-anti-e-neutrino attractive force because the energies of the anti-e-neutrinos are presumably much larger than the rest masses of the largely-non-relativistic electrons, and (2) the energy spectrum of the anti-e-neutrinos is unknown, and we try to avoid calculations in which too many unknowable assumptions must be made.

Suppose M is the mass of a spherically symmetric galaxy within a radius r and there exists a star with mass m located a distance r from the center of the galaxy. The combined Lee-Yang and gravitational forces on the star by the galaxy are

$$F_{com} = F_{YM} + F_{LY}, \tag{2.28}$$

$$F_{YM} = -\frac{GmME_\nu}{r^2 m_p}(1+\beta^2), \qquad F_{LY} = -\frac{g_\ell^2 mME_\nu}{4\pi r^2 m_p^2 m_\nu},$$

where we have used (2.27) and the first equation in (2.12) for concreteness. Because the mass of a galaxy is mainly due to its protons (we ignore neutrons in this estimation since 75% of the cosmic mass is hydrogen), M/m_p ($m_p = 0.94 GeV$) is the approximate number of protons in the galaxy within a radius r. Under the assumptions of the HEN model, M/m_p also represents the number of electrons and anti-e-neutrinos within a radius r. The resulting leptonic charge within radius r is $g_\ell M/m_p$ due to the electrons and $-g_\ell M/m_p$ due to the anti-e-neutrinos. Similarly, m/m_p is the number of protons and electrons in the star. Naturally, there is also a repulsive Lee-Yang force between the electrons in the star and the electrons within a radius r of the center of the galaxy. However, because the galactic electrons are virtually all non-relativistic, unlike the anti-e-neutrinos, the energy enhancement factor (E/m) in the Lee-Yang force Eq. (2.27) insures that this repulsive electron-electron Lee-Yang interaction is negligible compared to the attractive electron-anti-e-neutrino Lee-Yang interaction.

From observed rotation curves, the mass of the dark matter in a galaxy is estimated to be $M_D \approx N_D M$ where N_D has a value roughly between 5 and 60.[11,12] Since this dark matter is assumed to only interact gravitationally, the gravitational force between the galactic dark matter with mass M_D located within a distance r from the center of a galaxy and our star is

$$F^{grav} = -\frac{GmM_D}{r^2}.$$
(2.29)

To produce the observed flat rotation curve, the combined forces in (2.28) must be equal to the corresponding effective gravitational force (2.29),

$$\frac{GmME_\nu}{r^2 m_p}(1+\beta^2) + \left(\frac{g_\ell^2 mM}{m_p^2}\right)\left(\frac{E_\nu}{m_\nu}\right)\frac{1}{4\pi r^2} \approx \frac{Gm(N_D M)}{r^2},$$
(2.30)

$$or \qquad (1+\beta^2)\frac{E_\nu}{m_p} + \frac{g_\ell^2 E_\nu}{4\pi G m_p^2 m_\nu} \approx N_D.$$
(2.31)

Figure 2.1 shows the relative strength of the Yang-Mills gravitational force and the Lee-Yang leptonic force as a function of the anti-e-neutrino energy for the two different approximations for the static solutions discussed previously (Eqs. (2.5)–(2.12)). Of course, in reality, the anti-e-neutrinos will have a wide spectrum of energies, but the following plots serve to help us understand when one force or the other may be dominant. The lines on the two graphs correspond to the extremes of the estimated dark matter mass.

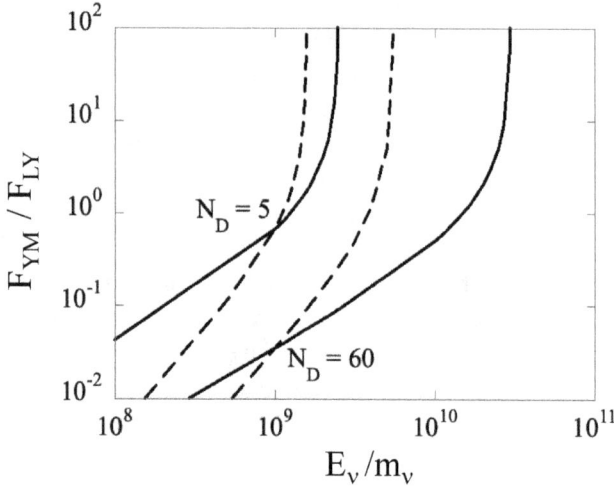

Fig. 2.1. Ratio of the Yang-Mills gravitational to Lee-Yang leptonic forces as a function of neutrino energy based on a flat galactic rotation curve. The solid and dashed lines correspond to the static approximations in Eq. (2.12A) and Eq. (2.12B), respectively.

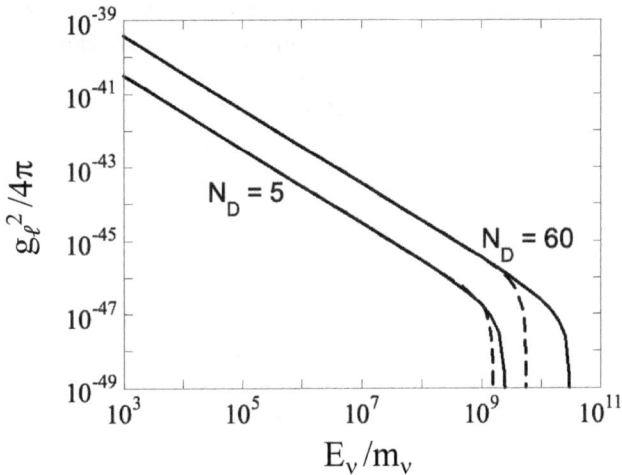

Fig. 2.2. Magnitude of the coupling constant for the Lee-Yang leptonic force as a function of neutrino energy based on a flat galactic rotation curve. The solid and dashed lines correspond to the static approximations in Eq. (2.12A) and Eq. (2.12B), respectively.

As one can see from Fig. 2.1, depending on N_D, the gravitational and leptonic forces are equal when E_ν/m_ν is roughly between 10^9 and 10^{10}. Furthermore, the results for the two different static approximations are similar. For less energetic anti-e-neutrinos, the flat rotation curve is principally due to the leptonic Lee-Yang force between the electrons in the star and the galactic anti-e-neutrinos, whose density distribution must be roughly $1/r^2$, consistent with current estimates. If the energies of the anti-e-neutrinos are, in general, higher than $10^9 m_\nu$, then the gravitational interaction between the electrons in the star and the galactic anti-e-neutrinos is the principal cause of the flat rotation curves. At very high anti-e-neutrino energies, the gravitational interaction is so large than there is no longer any real solution for the leptonic coupling constant g_ℓ. Figure 2.2 shows the inferred values for g_ℓ based on anti-e-neutrino energy and the observation of a flat galactic rotation curve.

2-6. The alpha magnetic spectrometer experiment and anti-e-neutrino dark matter

Because we expect the leptonic Lee-Yang force to be quite weak, it is most likely extremely difficult to devise an independent test of this aspect of the HEN model in an Earth-bound lab. However, anti-e-neutrinos $\bar{\nu}_e$ also interact via the weak interaction[c] in unified electroweak theory. This weak force has a much stronger coupling strength[3,4] ($\approx G_F m_p^2 = 10^{-5}$) than the leptonic Lee-Yang force or the gravitational force. Therefore, it is possible for anti-e-neutrinos to interact with cosmic-ray protons to produce positrons through the electroweak process,

$$\bar{\nu}_e + p^+ \rightarrow e^+ + n. \tag{2.32}$$

It turns out that the Alpha Magnetic Spectrometer (AMS) on the International Space Station has detected an intriguing excess of cosmic-ray positrons.[13] It is possible that this excess may be evidence of a much greater abundance of anti-e-neutrinos than is currently thought.

Let us briefly discuss the weak process (2.32) based on dimensional considerations and order-of-magnitude estimations. The total cross section of the process (2.32), when summed over all final states, is denoted by $\sigma(\bar{\nu}_e p^+)$. The amplitude of the weak process (2.32) should be proportional to the Fermi constant $G_F = 1.17 \times 10^{-5} GeV^{-2}$. Its total cross-section $\sigma(\bar{\nu}_e p^+)$ must have the form:[3]

$$\sigma(\bar{\nu}_e p^+) = G_F^2 f(s, m_p), \qquad G_F \approx 10^{-5} GeV^{-2}, \tag{2.33}$$

where s is the square of the center-of-mass energy. At high energies $s >> m_p^2$, the HEN model of anti-e-neutrino dark matter predicts

$$\sigma(\bar{\nu}_e p^+) \approx G_F^2 s. \tag{2.34}$$

If the prediction (2.34) of the HEN model could be tested and confirmed by the AMS experiment, it would support the interpretation of dark matter as being composed of high energy anti-e-neutrinos, as well as the postulate of the universal gauge symmetry.

We may compare the total cross-section (2.34) with neutrino-nucleon scattering in high energy laboratories,[8]

$$\sigma_{exp} \approx 10^{-38} \left(\frac{E_\nu}{m_p} \right) cm^2, \tag{2.35}$$

where the nucleon is at rest.[1]

---

[c]This interaction is effectively characterized by the Fermi constant[2] $G_F = 1.17 \times 10^{-5} GeV^{-2}$, which is much larger than Newton's constant $G = 6.7 \times 10^{-39} GeV^{-2}$.

2-7. Precision Cavendish experiments to test the leptonic Lee-Yang force

A second possible test of the cosmic Lee-Yang force[1] involves the force between two macroscopic bodies in a precision Cavendish experiment.[14, 15] Consider the Lee-Yang force F_{LY} between a pendulum ball with mass m_p and a source ball with mass m_s in a modern precision Cavendish experiment. For simplicity, we ignore the gravitational force due to the electrons and consider only the leptonic Lee-Yang force between the electrons of the two balls, which may be made of different materials. Suppose the pendulum ball with mass m_p is composed of an element with atomic number Z_p and atomic mass A_p, and the source ball with mass m_s is composed of an element with atomic number Z_s and atomic mass A_s.

The total number of the electrons in the source mass m_s is the product of the number of atoms in it, $m_s/(A_s a_u)$ (where a_u is the atomic mass unit) and its atomic number Z_s. Thus, the total leptonic charge carried by the electrons in the source mass is $g_{\ell s} = g_\ell m_s Z_s/(A_s a_u)$. Similarly, the total leptonic charge of the pendulum ball is $g_{\ell p} = g_\ell m_p Z_p/(A_p a_u)$. We note that the observed force F_{ob} between the source mass m_s and the pendulum ball with mass m_p is composed of both the repulsive Lee-Yang force $F_{LY} = g_{\ell p} g_{\ell s}/4\pi r^2$ and the attractive gravitational force F_g:

$$F_{ob} = -\frac{G_{eff} m_p m_s}{r^2} = F_g + F_{LY}, \qquad (2.36)$$

$$F_g = -\frac{G_g m_p m_s}{r^2}, \quad F_{LY} = \left(\frac{m_p Z_p}{A_p}\right)\left(\frac{m_s Z_s}{A_s}\right)\frac{g_\ell^2}{4\pi r^2 a_u^2},$$

$$a_u = 0.93149 \; GeV = 1.66054 \times 10^{-27} \; kg, \qquad c = \hbar = 1,$$

where the effective (or observed) Newtonian constant of gravitation G_{eff} differs from the true gravitational constant $G_g (\neq G_{ob})$ because of the Lee-Yang interaction. Based on experimental data, we know that the magnitude of the Lee-Yang force is smaller than that of the gravitational force because the relative standard uncertainty of the Newtonian constant of gravitation is 2.2×10^{-5} [Ref. 18], as deduced from various experiments, regardless of the materials used.

As a concrete example, let us use existing data from such an experiment to estimate g_ℓ. One measurement using a Cavendish-type apparatus with both balls made of copper with masses m_{p1} (pendulum) and m_{s1} (source) results in an effective value for the Newtonian constant of gravitation of $G_{eff1} = 6.67559U$,[14,15] where $U \equiv 10^{-11}m^3kg^{-1}s^{-2} = 1.0052 \times 10^{-39}GeV^{-2}$. From Eq. (2.36), we can write

$$(g_{\ell p}g_{\ell s})_{Cu} = \left(\frac{g_\ell^2 m_{p1}m_{s1}}{a_u^2}\right)\left(\frac{Z_{p1}Z_{s1}}{A_{p1}A_{s1}}\right)_{Cu}, \tag{2.37}$$

$$\left(\frac{Z_{p1}Z_{s1}}{A_{p1}A_{s1}}\right)_{Cu} = \left(\frac{29}{63.546}\right)^2 = 0.20824.$$

The same setup with a pendulum ball with mass m_{p2} made from quartz (SiO_2) and a source ball with mass m_{s2} made from iron, results in an effective value for the Newtonian constant of gravitation of $G_{eff2} = 6.67418U$.[16] For this case, we can write

$$(g_{\ell p})_{SiO_2}(g_{\ell s})_{Fe} = \left(\frac{g_\ell^2 m_{p2}m_{s2}}{a_u^2}\right)\left(\frac{Z_{p2}}{A_{p2}}\right)_{SiO_2}\left(\frac{Z_{s2}}{A_{s2}}\right)_{Fe}, \tag{2.38}$$

$$\left(\frac{Z_{p2}}{A_{p2}}\right)_{SiO_2}\left(\frac{Z_{s2}}{A_{s2}}\right)_{Fe} = \left(\frac{(14+16)}{(28+32)}\right)\left(\frac{26}{55.845}\right) = 0.23279.$$

Let us consider the difference of the following two quantities,

$$\frac{F_{ob1}}{(m_{p1}m_{s1})} - \frac{F_{ob2}}{(m_{p2}m_{s2})} = \frac{-G_{eff1}m_{p1}m_{s1}}{r^2(m_{p1}m_{s1})} + \frac{G_{eff2}m_{p1}m_{s2}}{r^2(m_{p2}m_{s2})}$$

$$= \frac{g_\ell^2}{4\pi r^2 a_u^2}\left[\left(\frac{Z_{p1}Z_{s1}}{A_{p1}A_{s1}}\right)_{Cu} - \left(\frac{Z_{p2}}{A_{p2}}\right)_{SiO_2}\left(\frac{Z_{s2}}{A_{s2}}\right)_{Fe}\right], \tag{2.39}$$

in which the true gravitational forces in $F_{ob1}/(m_{p1}m_{s1})$ (i.e., G_g/r^2) and in $F_{ob2}/(m_{p1}m_{s2})$ (i.e., G_g/r^2) cancel.

The strength of the leptonic charge $g_\ell^2/4\pi$ can be estimated from the observed difference of forces in (2.39). We obtain

$$\frac{g_\ell^2}{4\pi} = -a_u^2\left[G_{eff1} - G_{eff2}\right]Q \approx 5 \times 10^{-43}, \tag{2.40}$$

$$Q \equiv \left[\left(\frac{Z_{p1}Z_{s1}}{A_{p1}A_{s1}}\right)_{Cu} - \left(\frac{Z_{p2}}{A_{p2}}\right)_{SiO_2}\left(\frac{Z_{s2}}{A_{s2}}\right)_{Fe}\right]^{-1},$$

where we have used (2.37) and (2.38) and the experimental values of $G_{eff1} = 6.67559U$ and $G_{eff2} = 6.67418U$.[14,15] This estimate of g_ℓ in (2.40) gives us a feeling for the strength of the Lee-Yang force and is consistent with the value for g_ℓ corresponding to an anti-e-neutrino energy of roughly 1 MeV derived from the flat galactic rotation curves (see Fig. 2.2) in Sec. 2-5.

2-8. Summary

In summary, the HEN model represents an alternative approach to understanding dark matter phenomena that does not depend on the existence of as-yet undiscovered particles or other speculations. Instead:

(i) It provides an understanding of the flat galactic rotation curves without inventing new types of matter to drastically increasing the total observed mass of the universe by a factor of 5 to 60.

(ii) It is based on the well-established principle of gauge symmetry for quantum Yang-Mills gravity.

(iii) It is based on Broader Particle-Cosmology in a flat space-time with CPT invariance, implying the conservation of leptonic and baryonic charge, as well as an electrically neutral universe.

(iv) It assumes that high energy anti-e-neutrinos were produced through a dominant neutron decay process, $n \to p^+ + e^- + \bar{\nu}_e$, within the first 1000 seconds after the creation of quarks and antiquarks, at the beginning of the universe. This assumption is consistent with observations that there are almost equal number of protons and electrons in the observable cosmos.

(v) It predicts that the interaction between these ubiquitous anti-e-neutrinos and cosmic-ray protons leads to the excess of cosmic-ray positrons which have been detected by the Alpha Magnetic Spectrometer experiment. An excess of positrons in cosmic rays was also found by the HEAT balloon experiment by Barwick et al. and PAMELA.[13]

Furthermore, we know that dark matter particles must not interact significantly with radiation because dark matter has not lost sufficient kinetic energy to relax into the disk of galaxies, in contrast to baryonic matter.[17] The HEN model appears to be consistent with this property because the coupling constants associated with neutrinos are extremely weak, as shown in (2.33). These phenomena need to be further studied within the framework of Broader Particle-Cosmology with Yang-Mills gravity in flat space-time.

Although it is often stated that the dark matter problem is based on a well-confirmed theoretical construction, our present understanding of the universe is, in general, extremely limited. Thus, it is reasonable to investigate cosmology on the basis of Broader Particle-Cosmology, especially concerning phenomena from the early universe. Our discussions are based on the laws and properties established by theoretical particle physics which, in itself, comprise a well-confirmed theoretical framework. According to the present high energy neutrino (HEN) model, the effects of enormous mass of dark matter could be contributed from the high energies of anti-e-neutrinos (with enhanced gravitational and leptonic Lee-Yang forces) in our matter half-universe. Correspondingly, there are enormous masses of dark antimatter which are contributed from the high energies of the e-neutrinos in the antimatter half-universe. In view of the very weak couplings of neutrinos to other matter, the spectrum of these anti-e-neutrinos could give us information about the very early universe (something like the microwave background radiation). We hope that in the future there will be more data to test the HEN model.

References

1. J. P. Hsu, Chinese J. Phys. **74** 60 (2021) and references therein. See also M. Drees and F. Hajkarim, J. High Energy Phys. 1812, 042 (2018).
2. L. Landau and E. Lifshitz, *The Classical Theory of Fields* (tr. M. Hamermesh, Addison-Wesley, 1951) p. 89.
3. T. D. Lee, *Particle Physics and Introduction to Field Theory* (New York, Hardwood Academic, 1981) pp. 161–168, p. 177, pp. 210–211, pp. 320–333.
4. K. Huang, *Quarks Leptons & Gauge Fields* (World Scientific, 1982) p. 2, p. 6.
5. J. P. Hsu, Phys. Rev. Lett. **42**, 934 (1979). doi: 10.1103/PhysRevLett.42.934.
6. L. B. Okun, Phys. Today, The concept of mass, (June), **42**, 31 (1989).
7. J. P. Hsu and L. Hsu, *Space-Time, Yang-Mills Gravity, and Dynamics of Cosmic Expansion* (World Scientific, 2020) p. xxiii, pp. 106–110, p. 122, p. 133, pp. 135–137, pp. 212–213, pp. 266–269, pp. 289–294. F. J. Dyson, Bulletin of the American Math. Soc., 78, Sept. 1972. See also J. P. Hsu and D. Fine, *100 Years of Gravity and Accelerated Frames, The Deepest Insight of Einstein and Yang-Mills* (World Scientific, 2005) pp. 347–352, (with A Brief Remark for 'Missed Opportunity' by Dyson).
8. K. Nakamura et al., *Particle Physics Booklet*, pp. 137–138 (2010).
9. P. M. Ho, Generalized Yang-Mills theory and gravity, arXiv:1501. 05378v4 [hep-th], and references therein.
10. T. D. Lee and C. N. Yang, Phys. Rev. **98**, 1501 (1955), and references therein.
11. See, for example, K. G. Begeman, A. H. Broeils, R. H. Sanders, Monthly Notices of the Royal Astronomical Society, **249**, 523 (1991), (DOI: 10.1093/mnras/249.3.523); A. Riess, https://www.britannica.com/science/dark-matter. (accessed 28 Jan. 2021).
12. https://science.nasa.gov/astrophysics/focus-areas/what-is-dark-energy. (accessed 28 Jan. 2021).
13. CERN, Latest results from the AMS experiment, CERN, A. Lopes, (31 MAY 2018) (accessed 28 January 2021); M. Boezio, 'The PAMELA experiment: A cosmic ray experiment deep inside the heliosphere', 35th Int. Cosmic Ray Conf. (2017), DOI:10.22323/1.301.1091.
14. C. Rothleiner and S. Schlamminger, Rev. Sci. Instrument, 88, 111101 (2017); T. J. Quinn et al., Phys. Rev. Lett. **87**, 111101 (2001).

15. Qing Li et al., Nature **560**, 582 (2018); V. Milyukov, Proc. of the 9th Asia-Pacific Int. Conf. on Gravitation and Astrophysics, (Ed. J. Luo, Z. B. Zhou, H. C. Yeh and J. P. Hsu, World Scientific, 2010) pp. 3–15.

16. Qing Li et al., Nature **560**, 582 (2018).

17. S. Weinberg, *Cosmology* (Oxford Univ. Press, 2008) p. 39, pp. 185–200.

18. C. Rothleitner and S. Schlamminger (2017) Rev. of Sci. Instr. [online]; T. J. Quinn et al., Phys. Rev. Lett. 87, 111101 (2001).

3

A Model for Dark Energy

3-1. Baryonic charge and general Yang-Mills symmetry

The phenomenon of the late-time accelerated cosmic expansion is so different from our expectations that it seems very difficult to devise a reasonable explanation for it based on established physics. However, Broader Particle-Cosmology presents a possible avenue for understanding this phenomenon in terms of the baryon charges of protons and neutrons. Conventionally, one would consider the baryon charge to be associated with the conventional U_1 gauge symmetry, just like the leptonic or electric charges. However, Broader Particle-Cosmology allows the possibility that the baryon charge could instead be associated with a general Yang-Mills U_1 symmetry, whose gauge transformations involve an arbitrary vector gauge function rather than the conventional scalar gauge function. As we discussed in the previous chapter with leptons, although the conservation of baryon charge has been established experimentally, its corresponding gauge field B_μ has not been detected in high energy laboratories. This may be due to the extreme smallness of the baryonic charges in Earth-bound laboratory situations.

The key implication of associating baryonic charge with a general Yang-Mills U_1 symmetry is that one has a new gauge invariant field equation, which leads to a static linear potential (and thus a distance-independent force) between two point baryonic charges. This force between two baryons would be repulsive, while the force between a baryon and an anti-baryon would be attractive. The existence of such a force makes possible a situation where billions of years ago, when the galaxies were closer together, the attractive gravitational force was dominant, leading to a cosmic expansion taking place at a decreasing rate. However, when the distance between galaxies exceeded a certain critical distance, the repulsive baryon-baryon force became dominant, resulting in the accelerated expansion we now observe. Naturally, it must be the case that the baryon-baryon force is extremely small such that it has thus far evaded detection in the lab, becoming important only when cosmic-sized objects with enormous numbers of baryons, are involved. Because this force is distance-independent, the distance between two baryonic objects does not matter, only the number of baryons involved. (For details, see sections 3-3 and 3-4 below.)

Under the framework of Broader Particle-Cosmology, we make the

35

following proposal concerning the gauge symmetries of baryonic and leptonic charge:

(i) Leptons carry electric charge and leptonic charge. Both these charges are associated with the conventional U_1 gauge symmetry, whose transformations involve arbitrary scalar gauge functions and the usual phase functions. This produces $1/r$ (Coulomb-like) potentials. The resulting $1/r^2$ forces are responsible for electromagnetic phenomena, and may also be responsible for effects such as the flat galactic rotation curves and other dark matter-related effects through interactions with high energy anti-e-neutrinos, as discussed in the previous chapter.

(ii) Quarks carry electric charge, color charge, and baryonic charge. While the electric charge is associated with the conventional U_1 gauge symmetry, the color and baryonic charges are associated with general Yang-Mills symmetry, whose transformations involve arbitrary vector gauge functions and Hamilton's characteristic phase functions. This produces potentials that increase linearly with distance. The resulting distance-independent forces can explain phenomena such as the confinement of quarks (see Sec. 3 in Ch. 8) and the late-time accelerated cosmic expansion (to be discussed in this chapter).

We note that both cases can be treated under the same framework of general Yang-Mills symmetry. Although the general Yang-Mills U_1 symmetry involves vector gauge functions $\Lambda^\mu(x)$ and leads to fourth-order field equation and linear potentials, the conventional U_1 gauge symmetry is merely a special case of the general Yang-Mills U_1 symmetry where the vector gauge function $\Omega_\mu(x)$ can be expressed in terms of a scalar gauge function $\Omega(x)$, where $\Omega_\mu(x) = \partial_\mu\Omega(x)$. As a result, the leptonic gauge field equation is a second-order differential equation and results in a Coulomb-like potential. This situation where the usual gauge symmetry is a special case of a general Yang-Mills symmetry also holds for other symmetries such as SU_2 and SU_3.

3-2. Dynamics of baryonic and leptonic charges

We now consider the Lagrangians for the baryonic and leptonic charges. As we will see, it is necessary for the baryonic part of the Lagrangian to involve higher order derivatives, so that the baryonic gauge fields obey fourth-order differential equations.

In constructing the baryon-lepton (BL) Lagrangian, we note that historically, Gamba, Marshak and Okubo noted[1] a correspondence between baryons and leptons in the charged weak current under the interchanges $e \leftrightarrow n$, $\mu \leftrightarrow \Lambda$ and $\nu \leftrightarrow p$. After the discovery of quarks, this became known as the 'quark-lepton correspondence' where the correspondence was now between the six quarks and six leptons. Thus, we postulate the BL Lagrangian L_{BL} to have a quark-lepton correspondence under the exchanges:

$$q \leftrightarrow \ell, \quad m_q \leftrightarrow m_\ell, \tag{3.1}$$

$$q = (u, d, s, c, b, t), \quad \ell = (e, \nu_e, \mu, \nu_\mu, \tau, \nu_\tau). \tag{3.2}$$

An invariant Lagrangian that satisfies this condition and has different U_1 symmetries for the baryon and lepton gauge fields is

$$L_{BL} = +i\bar{q}\gamma^\mu(\partial_\mu + ig_b B_\mu)q - m_q\bar{q}q + \frac{L_b^2}{2}\partial^\mu B_{\mu\lambda}\partial_\nu B^{\nu\lambda} \tag{3.3}$$

$$+i\bar{\ell}\gamma^\mu(\partial_\mu + ig_\ell L_\mu)\ell - m_\ell\bar{\ell}\ell - \frac{1}{4}L_{\mu\lambda}L^{\mu\lambda},$$

where g_l is the lepton charge of a single lepton (an anti-lepton would have a charge of $-g_l$) and g_b is the baryon charge of a single quark (an anti-quark would have a charge of $-g_b$). Because the number of μ and τ leptons in the universe is negligible compared to the number of leptons from the electron family, we ignore them for cosmological applications. Thus, lepton number can be understood as the e-lepton number. Similarly, all quarks except for the u and d can be neglected.

The gauge curvatures $B_{\mu\nu}$ and $L_{\mu\nu}$, associated with $B_\mu(x)$ and $L_\mu(x)$, respectively, are defined by the commutators

$$[\Delta_{B\mu}, \Delta_{B\nu}] = ig_b B_{\mu\nu}, \quad [\Delta_{L\mu}, \Delta_{L\nu}] = ig_\ell L_{\mu\nu}, \tag{3.4}$$

where $\Delta_{B\mu}$ and $\Delta_{L\mu}$ are the usual gauge covariant derivatives

$$\Delta_{B\mu} = \partial_\mu + ig_b B_\mu(x) \quad and \quad \Delta_{L\mu} = \partial_\mu + ig_\ell L_\mu(x). \tag{3.5}$$

For leptons, we assume that the general Yang-Mills U_1 transformation is associated with a special vector gauge function $\Omega_\mu(x)$ that can be expressed in terms of a space-time derivative of an arbitrary scalar function $\Omega(x)$ at any space-time point x,

$$\Omega_\mu(x) = \partial_\mu\Omega(x), \tag{3.6}$$

making it equivalent to the conventional U_1 gauge symmetry.[2,3] Its general U_1 gauge transformation with the constraint (3.6), is then

$$L'_\mu(x) = L_\mu(x) + \Omega_\mu(x) = L_\mu(x) + \partial_\mu\Omega(x), \tag{3.7}$$

$$L'_{\mu\nu}(x) = \partial_\mu L'_\nu - \partial_\nu L'_\mu = L_{\mu\nu}(x);$$

$$\overline{\ell}'(x)\Delta'_{L\mu}\ell'(x) = \overline{\ell}'(x)\Omega_L(x)[\partial_\mu + ig_\ell L_\mu(x)]\ell(x) = \overline{\ell}(x)\Delta_{L\mu}\ell(x), \qquad (3.8)$$

$$\Omega_L(x) = e^{-iP_L}, \quad P_L = \left(g_\ell \int_{x'_o}^{x'_e = x} dx'^\mu \Omega_\mu(x')\right)_{Leb} = g_\ell \Omega(x), \qquad (3.9)$$

$$\ell'(x) = \Omega_L(x)\ell(x), \quad \overline{\ell}'(x) = \overline{\ell}(x)\Omega_L^{-1}(x), \qquad (3.10)$$

where $\Omega_L(x)$ in (3.8) and (3.9) can be shown to be the usual phase factor in the conventional U_1 gauge transformation. As usual, we assume that the starting point x'_o is fixed and does not contribute to P_L. The subscript 'Leb' in (3.9) (i.e., Hamilton's characteristic function P_L) denotes that the path of the integral in P_L must satisfy $[\partial_\mu \Omega_\lambda(x) - \partial_\lambda \Omega_\mu(x)]dx^\mu = 0$. This relation is automatically satisfied for arbitrary $\Omega_\mu(x)$ because of the constraint (3.6).

For baryons, we assume the more general Yang-Mills U_{1b} symmetry, whose gauge transformations involve general vector gauge functions $\Lambda_\lambda(x)^{\text{a}}$ and Hamilton's characteristic phase factor $\Omega_B(x)$,[4]

$$q'(x) = \Omega_B(x)q(x), \qquad \overline{q}'(x) = \overline{q}(x)\Omega_B^{-1}(x),$$

$$\Omega_B(x) = e^{-iP(x)}, \quad P(x) = \left(g_b \int_{x'_o}^{x'_e = x} dx'^\lambda \Lambda_\lambda(x')\right)_{Leb}, \qquad (3.11)$$

$$\delta P(x) = -ig_b \Lambda_\mu(x')\delta x'^\mu\big|_{x'_o}^x + ig_b \int_{x'_o}^x [\partial'_\mu \Lambda_\lambda(x') - \partial'_\lambda \Lambda_\mu(x')]dx'^\mu \delta x'^\lambda.$$

The subscript '*Leb*' in (3.11) denotes that the path of the action integration in $\delta P(x)$ must satisfy the equation,

$$[\partial_\mu \Lambda_\lambda(x) - \partial_\lambda \Lambda_\mu(x)]dx^\mu = 0. \qquad (3.12)$$

Thus, Hamilton's characteristic phase factor $\Omega_B(x)$ in (3.11) is unambiguously a local function of x. The variation $\delta\Omega_b(x)$ in (3.11) and the constraint (3.12) leads to the local relation at any space-time point x,

$$\frac{\partial \Omega_B(x)}{\partial x^\mu} = -ig_b \Lambda_\mu(x), \qquad (3.13)$$

which is important for the general U_{1b} symmetry. One can see that the usual coupling terms between the baryonic phase field $B_\mu(x)$ and the quark

---

[a]In general, the vector gauge function $\Lambda_\lambda(x)$ cannot be expressed in terms of a scalar function.

field $q(x)$ in (3.3) are invariant under new general U_1 gauge transformations involving vector gauge functions $\Lambda_\mu(x)$ as follows:

$$\partial^\mu B'_{\mu\nu}(x) = \partial^\mu B_{\mu\nu}(x) + \partial^\mu \partial_\mu \Lambda_\nu(x) - \partial^\mu \partial_\nu \Lambda_\mu(x) = \partial^\mu B_{\mu\nu}(x), \quad (3.14)$$

$$B'_{\mu\nu}(x) = B_{\mu\nu}(x) + \partial_\mu \Lambda_\nu(x) - \partial_\nu \Lambda_\mu(x) \neq B_{\mu\nu}(x),$$

$$constraint: \quad \partial^\mu \partial_\mu \Lambda_\nu(x) - \partial^\mu \partial_\nu \Lambda_\mu(x) = 0;$$

$$\bar{q}'(x)\gamma^\mu \Delta'_{B\mu} q'(x) = \bar{q}(x)\gamma^\mu \Delta_{B\mu} q(x), \quad (3.15)$$

$$\Delta_{B\mu} = \partial_\mu + ig_b B_\mu(x), \qquad B'_\mu(x) = B_\mu(x) + \Lambda_\mu(x). \quad (3.16)$$

Result (3.14) shows that the usual gauge curvature $B_{\mu\nu}$ is, in general, no longer invariant under the new general gauge transformations involving the vector gauge function Λ_μ. However, the divergence of the gauge curvature, i.e., $\partial^\mu B_{\mu\nu}$, in (3.14) can be invariant under the new gauge transformations, provided the constraint in the last equation in (3.14) is imposed. This constraint is automatically satisfied if $\Lambda_\mu(x) = \partial_\mu \Lambda(x)$, which is not true in the general U_1 gauge symmetry. We stress that the local relation (3.13) is crucial for the general Yang-Mills invariant coupling of quarks and the phase field $B_\mu(x)$, as shown in (3.15).

Thus, the new transformations for B_μ in (3.11)–(3.15) define the generalized U_1 gauge transformations for baryons. The corresponding vector fields B_μ are called 'phase fields,' which satisfy fourth-order equations rather than the usual second-order equations.

3-3. The distance-independent baryon-baryon force

Let us consider the Lagrangian (3.3)

$$L_{BL} = \frac{L_b^2}{2} \partial^\mu B_{\mu\lambda} \partial_\nu B^{\nu\lambda} - \frac{1}{4} L_{\mu\lambda} L^{\mu\lambda}$$

$$+ i\bar{q}\gamma^\mu(\partial_\mu + ig_b B_\mu)q - m_q \bar{q}q + i\bar{e}\gamma^\mu(\partial_\mu + ig_\ell L_\mu)e - m_e \bar{e}e + ..., \quad (3.17)$$

where we have ignored the neutrino terms.

The summation in the quark terms, such as $m_q\bar{q}q$, is understood. The function q in (3.17) denotes the Dirac spinor for a quark field with mass m_q. As mentioned earlier, in calculations of cosmological phenomena, we consider only u and d quarks, as well as electron-leptons.

The fourth-order equations of the general phase field B_μ can be derived from the Lagrangian (3.17)

$$L_b^2 \partial^2 \partial^\lambda B_{\lambda\mu} - g_b \bar{q}\gamma_\mu q = 0, \tag{3.18}$$

where L_b is a constant with the dimensions of length (i.e., a fundamental length) in the baryon terms of the Lagrangian (3.17) and the B_μ fields do not have self-coupling and are generated by quarks. If we impose the gauge condition $\partial^\lambda B_\lambda = 0$, (3.18) then leads to the field equation

$$\partial^2 \partial^2 B_\mu = \frac{g_b}{L_b^2}\bar{q}\gamma_\mu q. \tag{3.19}$$

Suppose there is a single quark at the origin. The zeroth component static gauge potential $B_0(\mathbf{r})$ satisfies the fourth-order equation,

$$\nabla^2 \nabla^2 B_0(\mathbf{r}) = \frac{g_b}{L_b^2}\delta^3(\mathbf{r}). \tag{3.20}$$

The potential $B_0(\mathbf{r})$[5,6] results in a repulsive force F_{qq} between two point-like quarks,

$$B_0(\mathbf{r}) = -\frac{g_b}{L_b^2}\frac{r}{8\pi}, \qquad \mathbf{F}_{qq} = -g_b\boldsymbol{\nabla}B_0 = \frac{1}{8\pi}\frac{g_b^2}{L_b^2}\frac{\mathbf{r}}{r}, \tag{3.21}$$

where we have used the following relation for generalized functions,[5]

$$\int_{-\infty}^{\infty}\frac{1}{(\mathbf{k}^2)^2}exp(i\mathbf{k}.\mathbf{r})d^3k = -\pi^2 r. \tag{3.22}$$

This force is independent of the distance between the two quarks. It is attractive between a quark and an anti-quark, and repulsive between two quarks or two anti-quarks. We will call this force the Okubo force in honor of Prof. Susumu Okubo for his endeavors. Since the Okubo force associated with conserved baryonic charges has not been detected in Earth-bound laboratories, it is presumably very small. However, if one considers cosmic-scale structures such as galaxies, then the Okubo force might be appreciable.[6–10]

3-4. Okubo forces on a cosmic scale

Although the values of g_b and L_b in (3.21) are unknown, we could roughly estimate them as follows. There is data that suggests that the rate of expansion of the universe was not always increasing and that the accelerated expansion began only about 4 billion years ago, when the universe was approximately 9 billion years old.[11,12] From this information, one could estimate the size of the universe at this critical point to be $V_{uc} \approx R_{uc}^3 \approx (9$ billion years $\times$ speed of light$)^3 \approx 6 \times 10^{77} m^3$.

Now consider two galaxies with equal masses m_g. Because the vast majority of the mass of a galaxy consists of hydrogen, virtually all of a galaxy's mass is baryonic (protons, in this case). The baryonic charge of each galaxy may thus be approximated as $(3g_b)m_g/m_p$, where m_p is the proton mass and g_b is the baryonic charge of a single quark. The critical distance d_c between two galaxies, i.e., the inter-galactic distance beyond which the repulsive baryon-baryon force becomes larger than the attractive gravitational force, can be found by setting the magnitudes of the gravitational and Okubo forces equal

$$|F_{gra}| = \frac{Gm_g^2}{d_c^2} \approx F_{Ok} \approx \frac{(9g_b^2 m_g^2)}{(8\pi L_b^2 m_p^2)}, \tag{3.23}$$

$$d_c \approx \left(\frac{Gm_p^2 8\pi L_b^2}{(3g_b)^2} \right)^{1/2}, \tag{3.24}$$

where we have treated the galaxies as point masses because the distance between galaxies is typically much larger than the size of a galaxy. We assume that the galaxies in the universe are uniformly distributed (consistent with the conventional cosmological principle, which approximates the universe as homogeneous and isotropic). When the average distance between galaxies is equal to the critical distance d_c, the volume of space per galaxy is approximately d_c^3 and the total volume V_{uc} of the universe is

$$V_{uc} \approx N_g d_c^3 = N_g \left(\frac{\sqrt{8\pi G L_b m_p}}{3g_b} \right)^3, \tag{3.25}$$

where N_g is the number of galaxies in the universe. Current estimates for the number of galaxies in the universe N_g vary, but a representative number is $N_g \approx 200$ billion. Setting this equal to our estimate for the size of the universe 4 billion years ago, we find that

$$g_b/L_b \approx 10^{-41} (meter^{-1}). \tag{3.26}$$

To find values for L_b and g_b individually, we note that the length L_b appears as a fundamental length in the Lagrangian (3.17) for baryonic dynamics. There is also a fundamental length[13] $L_s = 0.082 fm$ in the Lagrangian for the charmonium associated with the general Yang-Mills SU_3 symmetry (see Part II, Ch. 9, Eq. (9.19)). From the viewpoint of a total symmetry-unified model, including baryonic and color charges, it is perhaps reasonable to assume that the Lagrangians with the general Yang-Mills symmetry have only a single fundamental length, i.e.,

$$L_b = L_s \approx 0.082 fm. \tag{3.27}$$

This result, along with (3.26), give a value for the dimensionless baryonic charge g_b of

$$g_b \approx 8 \times 10^{-58}. \tag{3.28}$$

This value is an unimaginably small coupling constant in particle physics and the Okubo force can probably never be detected in Earth-bound laboratories. Nevertheless, it may have observable cosmological effects. This result suggests an interesting relation based on the principle of gauge invariance, that a microscopic physical quantity such as a basic length in quark confinement $L_s \approx 0.082 fm$ could be related to the baryon charge $g_b \approx 8 \times 10^{-58}$ in (3.28) for the super-macroscopic phenomenon of cosmic acceleration. This would provide a coherent big picture of the symmetry-unification of all interactions as suggested in the Broader Particle-Cosmology.

3-5. Summary

We summarize the features of the proposed model of dark energy phenomena as follows:

(A) The proposed model is based on Broader Particle-Cosmology with general Yang-Mills symmetry for conserved baryon charges. In particular, the basic principle of gauge symmetry in inertial frames is rooted in established physics rather than ad hoc assumptions. Conventional models assert that the universe is composed of 4% baryons, 20% dark matter, and 76% dark energy.[11] Based on Broader Particle-Cosmology, we have demonstrated in Ch. 2 that dark matter phenomena could be due to the enhanced Yang-Mills gravity of high energy anti-e-neutrinos. In this chapter, we have demonstrated that the dark energy phenomena could be generated by a cosmic Okubo force associated with well-established baryonic charges. Together, this eliminates the need to postulate the existence of heretofore unknown forms of matter or energy that make up 96% of the cosmos. As a bonus, it is gratifying that Broader Particle-Cosmology is based on flat space-time, in which the space-time coordinates in inertial frames have well-established, operational definitions, in contrast to the conventional models of dark energy based on general relativity, with its more complicated curved space time that is incompatible with quantum field theories.

(B) The model based on the cosmic Okubo force of baryon charges predicts the late-time cosmic acceleration to be

$$a_{cos}(r) = (F_{Ok} + F_{gra})/m_g$$

$$= \frac{9m_g g_b^2}{8\pi m_p^2 L_b^2} \left(1 - \frac{d_c^2}{r^2}\right) \approx 10^{-12} \left(1 - \frac{d_c^2}{r^2}\right) \frac{m}{s^2}, \qquad (3.29)$$

where we have used (3.23) and r is the average distance between two galaxies. These calculations are carried out in an inertial frame in flat space-time based on the framework of Broader Particle-Cosmology. It appears that there is no simple way to compare (3.29) with a value for the accelerated cosmic expansion based on expanding space in curved space-time.

(C) According to the proposed model, the acceleration of a galaxy is the combined result of the repulsive Okubo force and the attractive gravitational force on it by other galaxies. Thus, a main prediction of the model is that in the early universe, when the gravitational force was dominant, not only was the expansion occurring at a decreasing rate (decelerating), the rate of change of the deceleration was decreasing, since the gravitational force has a $1/r^2$ dependence. In the present universe, when the Okubo force is dominant (as evidenced by the accelerating expansion), the acceleration should be constant, since the Okubo force is independent of distance. We hope that these properties of the late-time cosmic acceleration can be tested in the future.

Appendix to Chapter 3.   Approximately distance-independent Okubo force between two galaxies

Let us consider the case of a galaxy S with mass m (modeled as a point) that is located a distance r away from a second galaxy with mass M, modeled as a sphere with uniform baryon charges and a radius R_o. Our calculation shows that the Okubo force between these two galaxies is approximately distance-independent.

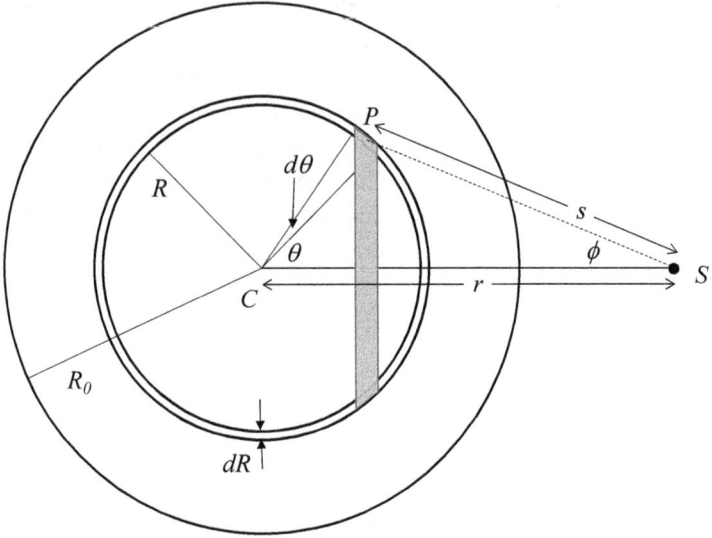

Fig. 3.1. A schematic diagram for calculations of the baryonic Okubo force between a point and a uniform sphere of baryonic charge.

To calculate the force on S by the uniform sphere,[7,8] we first divide the sphere into thin spherical shells with a radius $R < R_o$ and thickness dR. Now consider dividing each spherical shell into rings, as indicated by the shaded region in Fig. 3.1. The distance between S and every point of the ring is s and the angle PSC to be ϕ, as shown in Fig. 3.1. By symmetry, the baryonic force dF_{CS} on S by the ring must point along CS and, using (3.21), is given by

$$dF_{CS} = \int \frac{1}{8\pi L_b^2} \frac{3g_b m}{m_p} \frac{3g_b dM_r}{m_p} cos\phi, \qquad (3.\text{A}1)$$

where

$$dM_r = (\rho 2\pi R^2 sin\theta d\theta)dR. \qquad (3.\text{A}2)$$

The quantity dM_r is the mass of the ring where $2\pi R sin\theta$ is its circumference, $Rd\theta$ is its width, dR is its thickness, and ρ is the mass density of the sphere. In (3.A1), $3g_b m/m_p$ is the baryonic charge of S, $3g_b dM_r/m_p$ is the baryonic charge of the ring, and $cos\phi$ results from taking only the component of the force on S by the ring that is parallel to CS (by symmetry).

To evaluate this integral, we note that for the triangle CSP, we have

$$s^2 = R^2 + r^2 - 2Rr\,cos\theta, \qquad R^2 = s^2 + r^2 - 2rs\,cos\phi \qquad (3.\text{A}3)$$

by the law of cosines. The integration of $d\theta$ from 0 to π leads to an effective repulsive force on S along CS. The magnitude of the total repulsive force can be obtained[7] from the integration of ds:

$$dF_{CS} = A \int_0^\pi [dR\ R^2 cos\phi]\ sin\theta\ d\theta = A \int_{r-R}^{r+R} [dF] \frac{s}{Rr} ds, \qquad (3.A4)$$

$$A = \frac{(3g_b)^2 m\rho}{4L_b^2 m_p^2}, \quad dF \equiv dR\ R^2 \left(\frac{s^2 + r^2 - R^2}{2sr} \right). \qquad (3.A5)$$

Thus, the total repulsive baryonic force F^{Ok} exerted on the galaxy S with baryonic charge $+3g_b m/m_p$ by the galaxy modeled as a sphere with baryonic charge $3g_b M/m_p$ is

$$F^{Ok} = \int_0^{R_o} \frac{dF_{CS}}{dR}\ dR = \left(\frac{(3g_b)^2 mM}{8\pi L_b^2 m_p^2} \right) \left[1 - \frac{R_o^2}{5r^2} \right], \qquad (A3.6)$$

$$\approx \left(\frac{(3g_b)^2 mM}{8\pi L_b^2 m_p^2} \right), \quad R_o << r, \quad M = \frac{4\pi R_o^3 \rho}{3}. \qquad (3.A7)$$

When $r^2 >> R_o^2$, the cosmic Okubo force F^{Ok} is approximately a constant, i.e., independent of the distance r.

References

1. A. Gamba, R. E. Marshak, S. Okubo, Proc. Natl. Acad. Sci. **45** 881 (1951).
2. T. D. Lee and C. N. Yang, Phys. Rev. **98**, 1501, (1955). See also p. 211 and p. 600 in Ref. 10 below.
3. K. Huang, *Quarks, Leptons and Gauge Fields* (World Scientific, 1982), pp. 103–121 and pp. 241–242; see also Ref. 11.
4. W. Yourgrau and S. Mandelstam, *Variational Principle in Dynamics and Quantum Theory* (Dover, 3rd edition, 1979) p. 50; L. Landau and E. Lifshitz, *The Classical Theory of Fields* (Addison-Wesley, 1951) p. 29; S. Alahmad, Master Thesis, Univ. Massachusetts Dartmouth (2018).
5. I. M. Gel'fand and G. E. Shilov, *Generalized Functions* vol. 1 (Academic Press, New York, 1964) pp. 280–282, p. 363.
6. J. P. Hsu, Chin. Phys. C **41**, 015101 (2017). See also Alarmed in Ref. 4.
7. G. R. Fowler and G. L. Cassiday, *Analytic Mechanics* (Thomson, 2005) pp. 223–225.
8. J. P. Hsu and L. Hsu, *Space-Time, Yang-Mills Gravity and Dynamics of Cosmic Expansion* (World Scientific, 2020) pp. 268–269. See also Ref. 6.
9. J. P. Hsu, L. Hsu and D. Katz, Mod. Phys. Lett. A **33** 1850116 (2018).
10. T. D. Lee, *Particle Physics and Introduction to Field Theory* (Hardwood Academic, 1981) pp. 746–752.

11. J. A. Frieman, M. S. Turner, D. Huterer (2008), Ann. Rev. Astrono. and Astrophys. 46 (1): 385. arXiv:0803.0982. Dark Energy and the Accelerating Universe - arXivhttps://arxiv.org pdfPDF by J. Frieman (2008) Cited by 1818 arXiv:0803.0982v1 [astro-ph] 7 Mar 2008.

12. https://www.nasa.gov, Hubble Reaches New Milestone in Mystery of Universe's Expansion Rate. May 19, 2022.

13. E. Eichten, K. Gottfried, T. Kinoshita, J. Kogut, K. D. Lane, and T.-M. Yan, Phys. Rev. Lett. 34, 369 (1975).

4

The Question of the Missing Antimatter:
The Big Jets Model

4-1. The case of the missing antimatter

One big puzzling problem in our understanding of cosmology is the apparent absence of anti-particles in any significant quantity in the observable portion of the universe. Assuming that in its earliest stages, the universe was composed entirely of energy, the current state of affairs seems to run counter to our current understanding of particle physics. Within the framework of quantum field theory, the three general requirements of local quantum fields, Lorentz invariance and the spin-statistics relations for the quantization of fields, together imply the exact CPT invariance of any quantum field theory.[1]

All known Lorentz invariant field theories such as quantum electrodynamics, unified electroweak theory, quantum chromodynamics and quantum Yang-Mills gravity, have exact CPT invariance.[2] The symbol C (charge conjugate) denotes changing the sign of a charge, P (Parity, $\mathbf{r} \to -\mathbf{r}$) denotes space inversion and T denotes time reversal ($t \to -t$). If the Lagrangian of a field theory does not change under these three combined operations, then the theory is CPT invariant.

CPT invariance assures the exact lifetime and mass equalities between particles and their corresponding antiparticles.[1] It also implies opposite electroweak (and chromo-) interaction properties between particles and antiparticles, and the same gravitational interactions between particles and antiparticles (in the sense of local quantum Yang-Mills gravity based on flat space-time). We note that Einstein's theory of gravity does not have CPT invariance because it is not a Lorentz invariant quantum field theory in flat space-time. In other words, General Relativity is incompatible with quantum mechanics and cannot be quantized.[3] CPT invariance implies that there should be virtually equal numbers of particles and antiparticles in the universe.[a]

---

[a]There may be an extremely small difference in the numbers of particle and antiparticles due to a very weak violation of particle-antiparticle symmetry.[1,4] However, this cannot account for the observed asymmetry of the universe because the probability of producing such antiparticles is roughly proportional to the square of their coupling strength, i.e., (weak coupling/strong coupling)$^2 \approx (1/10^{12})$.

Two possible avenues for understanding the lack of observed antimatter would appear to be either (1) to modify our current understanding of particle physics or (2) to modify our understanding of the universe and its evolution, particularly near the beginning, when the conditions for the matter/antimatter asymmetry presumably were set. We would argue that the former of those avenues seems less promising, given the success of quantum field theories in predicting and explaining the many particle physics experiments performed over many decades.

In contrast, our knowledge of the universe and its evolution are comparatively limited. In particular, the foundations of current cosmological models are quite old. One modern cosmological model was first discussed in 1922 by Friedmann on the basis of general relativity.[5] Lemaitre proposed in 1927 that an expanding universe could be traced back to an original 'primeval atom'.[5] The first observational evidence for the expanding universe was obtained by Hubble in 1929 from analysis of cosmic redshifts.[5] In 1948, Gamow, Alpher, Bethe and Herman[5,6] discussed the evolution of the universe and the origin of chemical elements. In 1965, the discovery of the microwave background blackbody radiation by Penzias and Wilson[7] further lent support to the physical model of Alpher and Gamow. All these theoretical foundations[8] were laid before the advent of quarks with color SU_3 gauge symmetry and hence, may need some revision.

In this chapter, we propose a model of the beginning of the universe, based on Broader Particle-Cosmology, that can explain the apparent lack of antimatter in the observed universe. We also discuss experiments that could assess the viability of our proposed model.[2,9] For a discussion of the early universe to make sense, it is necessary to have a theoretical framework compatible with quantum field theories and particle physics with quarks. From this perspective, General Relativity is not sufficient since it is incompatible with the flat space-time of quantum mechanics and field theories. However, Broader Particle-Cosmology with Yang-Mills gravity does provide such a theoretical framework.

It has been realized that particle physics is critical to understanding the very early universe. There is little doubt that conditions at the beginning of the universe were similar to those inside a gigantic high-energy particle collider. Based on our present understanding of quarks, leptons, gauge bosons, etc., it is likely that the strong interaction processes of quark-antiquark production, confinement, scattering and decay largely dictated the evolution of the very early universe.[2,10] Those creation processes directly involving leptons, which have no strong interactions, would be negligible according to established particle physics.

4-2. Beginning of the universe: Big Bang or Big Jets?

Building on the analogy between the very early universe and conditions in high energy particle colliders, we note that a common occurrence in high energy particle collisions is the creation of jets of particles,[1] of which the simplest configuration is two oppositely directed jets. This has led us to postulate that the universe was created, not as a single Big Bang, but as two Big Jets.[2]

One may picture these two jets as two gigantic fireballs moving away from each other, carrying the same magnitude of linear momentum (for observers in the center-of-mass coordinate of the two jets). In each fireball, there was an enormous number of particles and antiparticles. Particle physics suggests that these were principally quarks and antiquarks in the very beginning, because of their non-zero masses and strong interactions.[11-13] After the processes of quark confinement to form resonance hadrons (with extremely short lifetimes), those then decayed into stable hadrons.

We postulate that one of these fireballs happened to be dominated by stable matter particles, forming the 'matter half-universe' in which we reside. CPT invariance implies that the other fireball must have been dominated by stable antiparticles, forming an 'antimatter half-universe'.[2] Note that the existence of both dark matter and dark antimatter is also implied by the CPT theorem, provided dark matter is made of known particles. To observers in each half-universe, the evolution process would have looked roughly similar to that of a hot big bang model. Thus, the Big Jets model preserves matter-antimatter symmetry according to the CPT invariance and explains the absence of antiparticles and antimatter galaxies in that part of the universe we can observe. In this sense, the experimental absence of antiparticles in our observable cosmos could be considered as indirect evidence for the Big Jets model.

In the Big Jets model, both the original matter fireball and the antimatter fireball expanded and cooled to form 2.7 K matter and antimatter blackbodies, respectively. Presumably, these two blackbodies are separated by an extremely large distance (probably much more than 10 billion light years) so that electromagnetic radiation emitted from anti-stars or anti-supernovae in the antimatter half-universe are too weak to be detected by us. Nevertheless, with a little bit of luck, the microwaves emitted from the antimatter blackbody as a whole could be detectable. If the matter half-universe is modeled as a spherical (or ellipsoid) blackbody, then the

Big Jets model predicts that the hemisphere of the matter blackbody that faces the antimatter half-universe should be slightly warmer than the opposite hemisphere (which should have a uniform temperature). This suggests a picture of a very small hemispheric anisotropy. It is noteworthy that Planck satellite data suggests a hemispheric asymmetry and cold spot in the cosmic microwave background.[9]

4-3. Anisotropy in the cosmic microwave background: A glimpse of the missing antimatter?

Observations of the cosmic microwave background (CMB) do indeed indicate that there is a 'dipole anisotropy' in the temperature on the order of 10^{-3}, suggesting a 'hemispheric anisotropy' on the order of 10^{-5} or smaller.[9,13] However, this anisotropy is conventionally attributed to the movement of the Earth relative to the CMB, rather than a true anisotropy in the CMB. Still, one weakness of this conventional interpretation is that it relies on a peculiar ad hoc transformation of the temperature that is inconsistent with every first principles attempt to formulate a relativistic thermodynamics. In this and the following sections, we elaborate on this weakness and propose an alternative formulation of the Planck law that does not have such a deficiency. Under such a formulation, the observed anisotropy may actually be a true anisotropy of the CMB, and not simply an artifact of Earth's motion.

The Planck distribution for a blackbody at rest in an inertial frame $F(t, x, y, z)$ has been well-established. The distribution can be written as

$$P(\omega, T) = \frac{1}{[exp(\omega/k_B T) - 1]}, \qquad \hbar = c = 1. \qquad (4.1)$$

In this connection, the temperature T is simply a parameter that characterizes the spectral distribution. In a different frame, the energies ω of the photons in the Planck distribution (4.1) will, of course, be different.

In the conventional analysis of the CMB, the Planck distribution in another inertial frame $F'(t', x', y', z')$ that is moving with a constant velocity V with respect to F is assumed to have the form:

$$P'(\omega', T') = \frac{1}{[exp(\omega'/k_B T'(\theta')) - 1]}, \qquad \omega' = k'_0, \qquad (4.2)$$

with

$$T'(\theta') \equiv \frac{\omega' T}{\omega} = \frac{T\sqrt{1 - V^2}}{(1 + V cos\theta')}, \qquad (4.3)$$

where θ' is the angle between V and the line of sight, and the relativistic Doppler effect [i.e., $\omega = \omega'(1 + V\cos\theta')/\sqrt{1 - V^2}$] of special relativity is assumed.[13]

This conventional formulation of the Planck law leads to the following interpretation of CMB observations: The CMB is assumed to be isotropic in some reference frame F. In an inertial frame F' corresponding to the Earth, the blackbody distribution $P'(\omega', T')$ is assumed to be given by $P'(\omega', T') = P(\omega, T)$. In order for this to be true, one must have a velocity-angle dependent temperature T',

$$\frac{\omega}{T} = \frac{\omega'}{T'(\theta')}, \quad or \quad \frac{T'(\theta') - T}{T} \approx -V\cos\theta', \qquad (4.4)$$

where θ' is the angle between the velocity V of the F' frame (or the earth) as measured in F and the line of sight (or the photon velocity). Accordingly, the 'dipole anisotropy' is interpreted as due to the motion of the Earth with a velocity $V \approx 370km/sec$ with respect to F.[14–17] This interpretation has been used by researchers analyzing the CMB using satellite data from the Cosmic Background Explorer (COBE), Wilkinson Microwave Anisotropy Probe (WMAP) and the Planck satellite experiments.[15]

However, the temperature transformation (4.3) is not based on any theoretical considerations. Instead, it is an ad hoc assumption made for the sole purpose of being able to write the Planck law in F' in the form shown in (4.2). Not only is this temperature transformation inconsistent with every theoretical attempt to create a relativistically consistent thermodynamics,[18–23] it also lacks the group properties necessary to be consistent with special relativity itself.[b]

The simplest way to see that an angle-dependent temperature, as defined by (4.3), cannot possibly be consistent with Lorentz invariance is to note that it cannot be made to satisfy the condition,

$$\frac{\omega}{T} = \frac{\omega'}{T'(\theta')} = \frac{\omega''}{T''(\theta'')}, \qquad (4.5)$$

for three or more reference frames. Furthermore, there appears to be no other experimental support for temperature as such a 'one-component' or non-scalar quantity. As we show in Appendix 3, condition (4.5) can only be

---

[b]In Einstein's complete works, the editors mentioned an interesting story about Einstein's investigation of the transformation property of the temperature. In 1907, Einstein gave the relation $T' = T\sqrt{1 - V^2}$. Around 1950, a physicist asked Einstein a question. Einstein revisited the problem and obtained a different answer, namely, $T' = T/\sqrt{1 - V^2}$. This interesting story reveals the subtlety of temperature in relativistic thermodynamics.

satisfied by constructing the temperature as a 4-vector T_μ where $T_\mu \propto k_\mu$ in all frames. However, none of the discussions of the conventional formulation appear to even consider this possibility.

Such an interpretation of the Planck distribution with an angle-dependent temperature in (4.2)–(4.3) may be related to the belief that the principle of relativity holds everywhere and for all phenomena, except when it involves the universe as a whole.[16,17] Cosmic phenomena are, in general, treated using general relativity which, as we have noted, is incompatible with special relativity, as well as all the modern schemes for a quantum-mechanical description of nature. Since the Planck distribution is rigorously derived on the basis of quantum mechanics (see comments of Dyson and Wigner in Ref. 3, Sec. 1-1, Ch. 1), we question whether this is a sound line of reasoning.

4-4. Relativistically invariant formulation of the Planck distribution

In this section, we present an alternative formulation of the Planck distribution that is not only Lorentz-Poincaré (LP) invariant, but also involves temperature transformations consistent with those that have been proposed previously.[18–23] Our alternative formulation is based on two observations:

(i) The quantity $k_B T$ can be made invariant under a Lorentz transformation.

(ii) A blackbody at rest in an inertial frame can be associated with a velocity 4-vector $U^\mu = dx^\mu/ds = (1,0,0,0)$ in that frame, where $ds^2 = dt^2 - dx^2 - dy^2 - dz^2$, $(c = \hbar = 1)$.

Regarding the first observation, one could assume that the temperature T and k_B are both scalar invariants, $T' = T = T_{inv}$ and $k'_B = k_B$. Alternatively, consistent with proposals by Einstein and Planck,[18,22] one could assume that $T' = T\sqrt{1 - V^2}$ and $K'_B = K_B/\sqrt{1 - V^2}$. Although having different values for $k'_B = k_B$ may seem strange, the universality of the Boltzmann constant has never been established in a relativistic framework.

To formulate the LP invariant Planck distribution, we generalize ω in the original Planck distribution $P(\omega, T)$ by replacing it with the product of the four-velocity of the thermo-system $U^\mu = dx^\mu/ds$ and the wave four-vector $k^\mu = (\omega, k_x, k_y, k_z)$,

$$\frac{1}{[exp(\omega/k_B T) - 1]} \rightarrow \frac{1}{[exp(k_\mu U^\mu)/(k_B T) - 1]}, \qquad K_B T = inv.$$

In the rest frame of the thermo-system, the four velocity of the thermo-system is $U^\mu = (1, 0, 0, 0)$ so that $k_\mu U^\mu = k_0 = \omega$. Thus, we replace ω with an invariant quantity that is identical to it in the rest frame of the thermo-system. Because the Planck distribution was formulated before the advent of special relativity, one does not necessarily expect it, as originally conceived, to be Lorentz-Poincaré invariant. Such a replacement is analogous to the replacement of the "mass" in the laws of conservation of energy and momentum with a quantity that is identical to the mass in the rest frame of the object during the establishment of special relativity.

Since $k_\mu U^\mu$ is the product of two four-vectors, it is LP invariant, making the ratio $k_\mu U^\mu / k_B T = k_0 / k_B T$ also LP invariant (since $k_B T$ is also invariant). Thus, in this new LP invariant formulation of Planck's distribution, one can write the generalized distribution law for blackbody radiation of CMB observed in any inertial frame F as

$$B(k_\mu U^\mu, T) = \frac{1}{[exp(k_\mu U^\mu)/(k_B T) - 1]}, \qquad k_\mu U^\mu = \eta_{\mu\nu} k^\mu U^\nu, \qquad (4.6)$$

$$U^\mu = \frac{dx^\mu}{ds}, \qquad \eta_{\mu\nu} U^\mu U^\nu = 1, \qquad \eta_{\mu\nu} = (1, -1, -1, -1), \qquad ds^2 = \eta_{\mu\nu} dx^\mu dx^\nu,$$

where we have used the notation $B(k_\mu U^\mu, T)$ for the blackbody radiation distribution to distinguish it from the conventional formulation of $P(\omega, T)$ in (4.1) and (4.2).

To see the transformation properties of $B(k_\mu U^\mu, T)$, we assume that a blackbody is at rest in F (i.e., $U^\mu = (1, 0, 0, 0)$) and located at the origin of F for simplicity. In another inertial frame F' moving with a relative velocity V along the $+x$ axis relative to F, the 4-velocity U'^μ of the blackbody is,

$$U'^\mu = \left(\frac{1}{\sqrt{1 - V^2}}, \frac{-V}{\sqrt{1 - V^2}}, 0, 0 \right), \qquad (4.7)$$

$$k'_\mu U'^\mu = k'_0 U'^0 + k'_1 U'^1 = k'_0 \frac{[1 + V cos\theta']}{\sqrt{1 - V^2}} = k_\mu U^\mu = k_0,$$

where $cos\theta' = -k'_1/k'_0$ and k_0 is a constant (independent of V) because the thermodynamic system (the blackbody) is at rest in the F fame. The relation between k_0 and k'_0 is just the relativistic Doppler shift. Thus, in the F' frame, the law for blackbody radiation takes the explicit Lorentz-Poincaré invariant form:[26]

$$B(k'_\mu U'^\mu, T) = \frac{1}{exp[k'_\mu U'^\mu/(k_B T)] - 1}. \qquad (4.8)$$

Let us now see what this new invariant formulation of the Planck distribution implies regarding the observed anisotropy in the CMB.

We stress that what is directly measured in experiments observing the CMB is the frequency spectrum W' (see Eq. (4.9) below and Fixsen et al. in Ref. 15), i.e., the power emitted per unit projected area of a blackbody at temperature T_{inv}, into a unit solid angle in the frequency interval ω and $\omega + d\omega$.

Since the frequency spectrum is directly related to the energy momentum tensor of the radiation, let us consider the energy-momentum tensor of photons[16] and the blackbody frequency spectra W and W' in inertial frames F and F'. The differential energy-momentum tensor $dT_{\mu\nu}$ is defined to be proportional to $k_\mu k_\nu$, whose coefficient (in the rest frame of the blackbody) is determined so that the integral of the (00) component, dT_{00}, over θ and ϕ should be $u(\nu)d\nu = 8\pi h\nu^3 d\nu/(c^3[exp(h\nu/k_B T) - 1])$ (in cgs units), where $u(\nu)$ is the spectral energy density (i.e., energy per unit volume per unit frequency interval). Thus, we have

$$dT_{\mu\nu} = \frac{1}{2\pi^3} k_\mu k_\nu B(\eta^{\lambda\sigma} k_\lambda U_\sigma, T_{inv})\delta(\eta^{\alpha\beta} k_\alpha k_\beta)d^4 k, \qquad (4.9)$$

$$dT_{00} = W d\omega d\Omega, \qquad W = \frac{\omega^3}{4\pi^3} \frac{1}{(e^{k\cdot U/k_B T} - 1)},$$

$$dT'_{00} = W' d\omega' d\Omega', \qquad W' = \frac{\omega'^3}{4\pi^3} \frac{1}{(e^{k'\cdot U'/k_B T} - 1)},$$

$$\frac{W'}{W} \approx (\omega'/\omega)^3 \approx 1 - 3V\cos\theta', \qquad d\Omega = \sin\theta d\theta d\phi,$$

where the integration over $k_0 > 0$ in $dT_{\mu\nu}$ is understood. We have also used the Doppler shift in the second equation in (4.7) and the invariant blackbody distribution in (4.6) and (4.8).

From these results, we conclude that even though the blackbody distributions measured from any frame will be identical (as indicated by $k'_\mu U'^\mu = k_\mu U^\mu$ in (4.7)), the frequency spectra W and W' (which is what is actually measured by satellite experiments) will be different. In this formulation, however, any difference is due to the fact that W and W' depend not only on the invariant blackbody distribution, but also on the velocity-angle dependence of the Doppler frequency shift, rather than any temperature difference arising from the relative motion of the frames. Thus, the observed anisotropy will have to be analyzed in a different way to determine whether or not there is a real anisotropy in the CMB temperature.

4-5. Re-interpreting CMB observations

In order to analyze CMB measurements made from the Earth, which is strictly speaking not an inertial frame, one should actually use an even more general formulation of the Planck distribution for reference frames undergoing orbital motion. The derivation, based on the same considerations as the one above (in Sec. 4.4), can be found in Appendix 2 to Ch. 4. However, the main result is this:

Based on the principle of limiting continuation of physical laws for general reference frames, both inertial and non-inertial,[27, 28] the rotational frequency shift ((4.A8) in Appendix 2) implies that the general-frame (GF) covariant blackbody distribution $B(P_{\mu\nu}k_r^\mu U_r^\nu, T)$ measured in the earth frame (revolving around the sun) is the same as the blackbody distribution $B(k'_\mu U'^\mu, T)$ in an inertial frame F', i.e.,

$$B(P_{\mu\nu}k_r^\mu U_r^\nu, T) = B(k'_\mu U'^\mu, T). \tag{4.10}$$

Therefore, the GF covariant Planck distribution in (4.10) does not depend on the time w_r of the rotating frame, as shown in (4.A9) using the relation (4.A8) in Appendix 2. The frequency spectrum, however, does depend on the time w_r and will show a 6-month variation in the measured radiation energy k_0 in the Earth frame due to Earth's motion around the sun. The observed frequency spectra W' (in inertial frame F') and W_r (in a rotating frame F_r) are related to (4.9) by:

$$dT'_{00} = W' d\omega' d\Omega', \qquad dT_{r00} = W_r d\omega_r d\Omega_r. \tag{4.11}$$

The integration over $k'_0 > 0$ in $dT'_{\mu\nu}$ is carried out to obtain dT'_{00} in (4.11), where we have $\omega' = |\mathbf{k}'|$. The same thing is done for dT_{r00} in the rotating frame F_r. In the F' frame, W' is the power emitted per unit projected area of a blackbody at temperature T, into a unit solid angle, in the angular frequency interval from ω' to $\omega' + d\omega'$.

For simplicity, suppose the microwave receiver is located in the rotating frame F_r at $(x_r, y_r, 0) = (\rho, 0, 0)$. Using (4.11) for a rotating F_r frame and inertial F' frame to the first order in earth's orbital velocity $\Omega\rho$, we predict the ratio of the frequency spectrum W_r/W' to be time-dependent,

$$\frac{W_r}{W'} = \left(\frac{\omega_r}{\omega'}\right)^3 \approx 1 + 3\Omega\rho\,sin(\Omega w_r), \tag{4.12}$$

where the frequency is $\omega_r = k_{r0}$, and $w_r (\equiv \mathrm{w}_r)$ is the time in the rotating $F_r(w_r, x_r, y_r, z_r)$ frame. We have also used the rotational Doppler shift (4.A7) (in Appendix 2) and the invariant distribution in (4.10).

To be even more realistic, suppose the sun (or the F' frame) is moving with a constant velocity relative to the axis x'' of another inertial frame F'', which is at rest relative to CMB. Since the conventional interpretation of the CMB says that the velocity of the Earth is pointed towards the stars forming the Virgo cluster, we assume that the x''-axis is pointing toward the Virgo cluster. In this case, the frequency spectrum measured on the earth will be modified by both the Doppler effect due to Earth's orbital motion around the sun (W_r/W') and the inertial motion of the solar system through space (W'/W''). To the lowest order, we have

$$\left(\frac{W_r}{W'}\right)\left(\frac{W'}{W''}\right) \approx \left(\frac{\omega_r}{\omega'}\right)^3_{ro}\left(\frac{\omega'}{\omega''}\right)^3_{in} \tag{4.13}$$

$$\approx 1 + 3V + 3\Omega\rho \; sin(\Omega w_r).$$

The frequency shift at four values of Ωw_r are particularly interesting:

Ωw_r :	0	$\pi/2$	π	$3\pi/2$
velocity :	$3V$	$3V + 3\Omega\rho$	$3V$	$3V - 3\Omega\rho$

where $V \approx 120km/s$ and $\Omega\rho \approx 30km/s$ (in SI units). Therefore, the general-frame (GF) covariant law for blackbody radiation predicts that the blackbody energy spectrum W measured from an Earth orbiting the sun will a display a 6-month variation of the velocity toward Virgo, as shown above from a minimum velocity of $3V - 3\Omega\rho = 270km/s$ to a maximum velocity of $3V + 3\Omega\rho = 450km/s$, in SI units. By analyzing the CMB observed from Earth in this manner, one should be able to determine the energy spectrum of the CMB, and thus the temperature profile of the CMB from the rest frame of the CMB.

If such an analysis were to show that the temperature profile of the CMB were anisotropic, then the problem arises of interpreting this anisotropy. One possibility is that the temperature profile of the CMB is actually isotropic, but that the Planck distribution (4.10) is not actually invariant, i.e., that Lorentz-Poincaré invariance is violated. A second possibility is that the temperature profile of the CMB is actually isotropic and the thermodynamic temperature transforms in a way that we have not expected; in fact, it transforms in such a way as to fit the observed anisotropy. However, we would prefer to believe in a third possibility, that Lorentz-Poincaré invariance holds for the invariant blackbody distribution (4.10) and that the temperature profile of the CMB is truly anisotropic, with a physical cause. That physical cause may be an anisotropy in the distribution of matter outside the presently observed portion of the cosmos, such as the existence of a far away antimatter half-universe.

4-6. Discussion and summary

Based on inertial and non-inertial frames with the space-time transformations discussed in Ch. 1, we have formulated a general-frame covariant Planck law (4.10) in flat space-time, where all space and time measurements have a realizable, operational meaning.

The conventional analysis of CMB experiments is based on an ad hoc velocity-angle dependent temperature $T'(\theta') = T\sqrt{1 - V_c^2}/(1 + V_c cos\theta')$ with $V = V_c$ and a non-invariant Planck distribution $P'(\omega', T')$ in (4.2). In this interpretation, the observed CMB dipole anisotropy is due to the transformation properties of the temperature $T'/T \approx 1 - V_c cos\theta'$ (or $(T' - T)/T \approx -V cos\theta'$, $V = V_c$ in (4.4)), which is interpreted as resulting from the motion of the solar system with a velocity $V_c \approx 370 km/s$ (in SI units) in the direction of the Virgo cluster.[16]

In contrast, based on the Lorentz-Poincaré invariant blackbody distribution $B(\eta^{\lambda\sigma} k_\lambda U_\sigma, T_{inv})$ and the non-invariant frequency spectrum W', we find

$$\frac{W'}{W} \approx \frac{\omega'^3}{\omega^3} \approx 1 - 3V cos\theta' \qquad \leftrightarrow \qquad \frac{T'(\theta')}{T} \approx 1 - V_c cos\theta', \qquad (4.14)$$

$$V \approx V_c/3 \approx 120 km/s, \qquad \leftrightarrow \qquad V_c \approx 370 km/s,$$

where V and V_c are in SI units. The results for $T'(\theta')/T$ and V_c in (4.14) are for comparison of our results with the conventional interpretation. From this perspective, the observed CMB dipole-anisotropy of order $V_c/c \approx 10^{-3}$ is caused by the Doppler shift of the frequency spectrum W' rather than the velocity-angle dependence of a temperature transformation in the Planck distribution in (4.2). Again, the physical cause is the motion of the solar system in the direction of the Virgo cluster with a velocity $\approx 120 km/s$.

The usual experimental description of anisotropies in the CMB is, in general, based on an angle-dependent temperature expanded in terms of spherical harmonics,[15]

$$T(\theta, \phi) = \sum_{\ell m} a_{\ell m} Y_{\ell m}(\theta, \phi). \qquad (4.15)$$

Since the angle-dependent temperature does not have proper transformation properties, it is difficult to compare it with data. From the view point of the invariant Planck distribution formulation, it is better to use the frequency spectrum W rather than the temperature T,

$$W = \frac{\omega^3}{4\pi^3} \frac{1}{(e^{k \cdot U/k_B T} - 1)}, \qquad (4.16)$$

to analyze CMB measurements, where the frequency ω and the Planck distribution $1/(e^{k \cdot U/k_B T} - 1)$ have the proper transformation properties for inertial and non-inertial frames, as we have discussed. The frequency spectrum W is more directly related to experimental measurements.

A full analysis of the precise nature of the CMB anisotropy is beyond the scope of this book. However, if transforming the observed CMB using the invariant Planck distribution as discussed here does not result in an isotropic CMB in the rest frame of the CMB, then one must search for physical causes for any anisotropies found. Possible explanations will no doubt depend on the precise nature of the anisotropy. For example, a hemispheric anisotropy might be indicative of a distant antimatter half-universe. Thus, we advocate for a re-interpretation of the CMB data, using our Lorentz-Poincaré invariant formulation and transforming the observed frequency spectra using the Doppler shift, rather than a relativistically inconsistent temperature transformation. The Big Jets model implies the existence of a 'hemispheric anisotropy' in the rest frame of the CMB on the order of 10^{-5} or smaller.[2] Such a prediction could be tested by CMB data from the Planck satellite.[9] The significance of testing the antimatter half-universe, as predicted by the CPT invariance of particle physics, cannot be emphasized strongly enough.

Appendix 1 to Chapter 4.   New General-Frame Covariant Law for Blackbody Radiation in Inertial and Non-Inertial Frames

For the best use of satellite measurements of the CMB to test physical laws related to cosmic phenomena, let us first discuss the big picture of the laws of physics based on the general space-time framework, including inertial and non-inertial frames. In particular, reference frames with linear acceleration and rotating reference frames with operationally defined space and time coordinates are particularly interesting and relevant. Even though non-inertial frames are not equivalent to inertial frames and we do not have the principle of relativity to guide our understanding of physics, we can generalize the principle of relativity for inertial frames to the principle of limiting continuation of physical laws for both inertial and non-inertial frames as follows:

The laws of physics in a reference frame F_1 with an acceleration $\mathbf{a}_1$ must reduce to those in a reference frame F_2 with an acceleration $\mathbf{a}_2$ in the limit where $\mathbf{a}_1$ approaches $\mathbf{a}_2$.

Accordingly, when $\mathbf{a_2} = 0$, this principle of limiting continuation reduces to the usual principle of relativity in the limit of zero acceleration. In previous work, we have discussed a fundamental space-time framework that encompasses all inertial and non-inertial frames of reference in flat space-time.[28] To make physical predictions, we must have coordinate transformations between inertial frames and non-inertial frames. (See also discussions in Ch. 1.) Based on this space-time framework, the differential of the covariant energy-momentum tensor (4.16) for photons as well as other laws of physics in inertial frames can be naturally generalized to non-inertial frames with more complicated Poincaré metric tensors $P_{\mu\nu}$, which reduce to the Minkowski metric tensor $\eta_{\mu\nu}$ in the limit of zero acceleration.[28]

We shall use the term 'general-frame covariant' (GF covariant) to describe the property that a physical law in a non-inertial frame in flat space-time[c] is 'formally' similar to the corresponding law in an inertial frame when the Minkowski metric tensor $\eta_{\mu\nu}$ is replaced by the Poincaré metric tensor $P_{\mu\nu}(x)$ of the non-inertial frame and when ordinary derivatives by are replaced by covariant derivatives. This appears to be natural because the Poincaré metric tensor of a non-inertial frame reduces to the Minkowski metric tensor in the limit of zero acceleration.[d]

In a general non-inertial frame with the Poincaré metric tensor $P^{\lambda\nu}(x)$ in flat space-time, we postulate the differential of the GF covariant energy-momentum tensor and the Planck law for photons or blackbody radiation as follows,

$$dT_{\mu\nu} = \frac{1}{2\pi^3} k_\mu k_\nu B(P^{\lambda\sigma} k_\lambda U_\sigma, T_{inv}) \delta(P^{\alpha\beta} k_\alpha k_\beta) \sqrt{-det P^{\lambda\sigma}} d^4 k, \quad (4.\text{A}1)$$

$$\mathbf{B(k_\lambda U^\lambda, T_{inv})} = \frac{1}{[\exp(\mathbf{k_\mu U^\mu})/(\mathbf{k_B T_{inv}}) - 1]}, \quad (4.\text{A}2)$$

where $k_\lambda U^\lambda = P^{\lambda\nu} k_\lambda U_\nu$ and $P^{\lambda\nu}$ in different frames is as follows:
(A) In an inertial frame $F(w, x, y, z)$, we have

$$P^{\mu\nu} = \eta^{\mu\nu} = (1, -1, -1, -1).$$

---

[c]This differs from the concept of general coordinate invariance in general relativity based on curved space-time. In general relativity, the space-time coordinates are local and have no operational meaning, in contrast to the coordinates in special relativity based on flat space-time. Mathematically, the coordinates in the Riemann manifold are local and have no meaning.

[d]Mathematically, if a differential equation in inertial frames is known, the corresponding equation in non-inertial frames can be obtained simply by replacing ordinary derivatives by covariant derivatives.[30]

(B) In an orbiting/rotating frame $F_r(w_r, x_r, y_r, z_r)$, we have

$$P^{00} = \gamma^{-2}\left[1 - \Omega^4 w_r^2(x_r^2 + y_r^2)\right] \approx 1, \quad P^{11} \approx P^{22} \approx -1 + O(\Omega^2),$$

$$P^{33} = -1, \quad P^{01} \approx \Omega y_r, \quad P^{02} \approx -\Omega x_r, \quad \sqrt{-det\ P^{\lambda\sigma}} \approx 1.$$

(C) In a frame with a constant-linear-acceleration $F_{CLA}(w, x, y, z)$ with acceleration α_o, we have

$$P^{00} = [\gamma^2(\gamma_o^{-2} + \alpha_o x)]^{-2}, \quad P^{11} = P^{22} = P^{33} = -1, \quad \sqrt{-det\ P^{\lambda\sigma}} = \sqrt{P^{00}}.$$

(D) In a frame with an arbitrary-linear-acceleration $F_{ALA}(w, x, y, z)$ with acceleration $\alpha(w)$ and 'jerk' $J_e(w) = d\alpha(w)/dw$, we have

$$P^{00} = [\gamma^2(\gamma_o^{-2} + \alpha(w)x)]^{-2}, \quad P^{01} = \frac{J_e(w)}{\alpha^2(w)\gamma_o^2}, \quad P^{22} = P^{33} = -1,$$

$$P^{11} = -[\gamma^2(\gamma_o^{-2} + \alpha(w)x)]^{-2}\left[\gamma^2(\gamma_o^{-2} + \alpha(w)x) - \frac{J_e(w)}{\alpha^2(w)\gamma_o^2}\right]^2,$$

where other components of the symmetric tensor $P^{\mu\nu}$ vanish,[e] and the integration over $k_0 > 0$ in (4.A1) is understood. We also have $P^{\lambda\nu}k_\lambda k_\nu = \eta_{\mu\nu}k'^\mu k'^\nu = 0$ for photons, which leads to an accelerated Wu-Doppler shift for case (B). The GF covariant blackbody distribution (4.A2) is a general Planck distribution for non-inertial frames with the Poincaré metric tensor $P^{\mu\nu}$, as given above. In the limit of zero acceleration, we have $P_{\mu\nu} \to \eta_{\mu\nu}$, as required by the principle of limiting continuation of physical laws for general reference frames. Note that k_μ is the covariant wave 4-vector as measured in the non-inertial frame and the temperature T is assumed to be an invariant, $T = T_{inv}$ in inertial and non-inertial frames.[f]

Appendix 2 to Chapter 4.   General-Frame Covariant Blackbody Distributions in Non-Inertial Frames

To obtain a space-time coordinate transformation between an inertial frame and an orbiting/rotating frame, we note that the invariant form of the laws (4.A1) with the zero acceleration limit, i.e. $P^{\mu\nu} \to \eta^{\mu\nu}$, holds in any four-dimensional symmetry framework based solely on the first postulate of

---

[e] See Ch. 3 in Ref. 35.
[f] It suffices to assume $(k_B T)$ in (4.18) to be invariant. We simply assume invariant T for simplicity.

relativity.[27,28] In other words, the coordinates of the symmetry framework are denoted by (w, x, y, z) and (w', x', y', z'), where the evolution variables w and w' are measured in the unit of length. Thus, the usual universality of the speed of light $c = 299792458m/s$ expressed in units of meters per second is not necessary for the invariance of the Planck law and other physical laws in general. For simplicity, let us concentrate on the invariant distribution (4.A2) for an orbiting/rotating frame $F_r(w_r, x_r, y_r, z_r)$, in which the speed of light is not constant and w_r is expressed in units of length.[27,28] If one wishes, one can replace (w_r, x_r, y_r, z_r) by $(x_r^0, x_r^1, x_r^2, x_r^3)$ without affecting any physics.

Let us use x'^μ and x_r^μ respectively for the coordinates of an inertial frame F' and an orbiting/rotating frame F_r. The origin of F_r orbits the origin of F' at a distance R, with a constant angular velocity Ω. A Cartesian coordinate system is used in both frames, set up in such a way that the positive portion of the y_r-axis of F_r always extends through the origin of F'. The principle of limiting continuation of physical laws requires that the space-time coordinate transformations between F_r and F' must reduce to the Lorentz transformations in the limit of zero acceleration. These transformations have been discussed in detail elsewhere,[28] so we simply give the result here.

The generalized space-time transformations between $F'(w', x', y', z')$ and a rotating frame $F_r(w_r, x_r, y_r, z_r)$ are

$$w' = \gamma(w_r + \boldsymbol{\rho} \cdot \boldsymbol{\beta}), \qquad x' = \gamma[x_r \, cos(\Omega w_r) - (y_r - R) \, sin(\Omega w_r)],$$

$$y' = \gamma[x_r \, sin(\Omega w_r) + (y_r - R) \, cos(\Omega w_r)], \qquad z' = z_r, \qquad (4.A3)$$

$$\beta = \Omega\sqrt{x_r^2 + (y_r - R)^2} = \Omega S < 1, \qquad \boldsymbol{\rho} \cdot \boldsymbol{\beta} = x_r R\Omega, \qquad (4.A4)$$

$$\gamma = \frac{1}{\sqrt{1 - \beta^2}}.$$

They reduce to the Lorentz transformation in the limit $R \to \infty$ and $\Omega \to 0$ such that their product $R\Omega = \beta_o$ is a non-zero constant velocity parallel to the x' axis: $w' = \gamma_o(w_r + \beta_o x_r), x' = \gamma_o(x_r + \beta_o w_r)$, etc.[28,29]

The resultant rotational transformations (4.19) are exact and consistent with the precision laser experiments of Davies-Jennison and Thim.[28, 29] They are also consistent with the results of high energy experiments involving unstable particles in a circular storage ring, and support the analysis of the Wilson experiment.[25]

To see the implications of the General-Frame covariant blackbody (or Planck) distribution (4.A2) for a rotating frame, let us approximate the Earth's motion around the sun as circular with radius ρ, at a constant angular velocity Ω which is, by definition, measured in the rotating frame F_r.[29] In this case we may set $R = 0$ in (4.A3) and (4.A4) for simplicity and the origins of both frames at the sun.

The covariant wave 4-vector for blackbody radiation is denoted by $k'_\mu = (k'_0, k'_1, 0, 0)$, as measured in F'. An Earth-bound observer (or the microwave receiver) is at rest in the rotating frame F_r and located at $(x_r, y_r, 0)$. For simplicity, let us ignore the relative linear motion between the solar system frame F' and the CMB rest frame, and concentrate on the effect of Earth's orbital motion (i.e., the 6-month variation) on the generalized distribution (4.A2). To first order in $x_r\Omega$ and $y_r\Omega$ for the Poincaré metric tensors $P_{\mu\nu}(x)$, we have[29]

$$P^{00} = \gamma^{-2}\left[1 - \Omega^4 w_r^2(x_r^2 + y_r^2)\right] \approx 1, \quad P^{11} \approx P^{22} \approx -1, \quad P^{33} = -1,$$

$$P^{01} \approx \Omega y_r, \quad P^{02} \approx -\Omega x_r, \quad \sqrt{-\det P^{\lambda\sigma}} \approx 1. \tag{4.A5}$$

To evaluate $P_{\mu\nu}k_r^\mu U_r^\nu = k_{r\nu}U_r^\nu$, we may use $U'^\mu = dx'^\mu/ds$, $U_r^\mu = dx_r^\mu/ds$ and $U'^\mu = (1,0,0,0)$ to solve for $U_r^\nu = (U_r^0, U_r^1, U_r^2, 0)$, where we assume the blackbody to be at rest in the inertial frame F'. To the first order in Ωx_r and Ωy_r, we obtain

$$U'^0 \approx U_r^0 \approx 1,$$

$$U'^1 \approx U_r^1 \cos\Omega w_r - U_r^2 \sin\Omega w_r - (\Omega x_r \sin\Omega w_r + \Omega y_r \cos\Omega w_r) \approx 0,$$

$$U'^2 \approx U_r^1 \sin\Omega w_r + U_r^2 \cos\Omega w_r + (\Omega x_r \cos\Omega w_r - \Omega y_r \sin\Omega w_r) \approx 0$$

which lead to

$$U_r^1 \approx \Omega y_r, \quad U_r^2 \approx -\Omega x_r.$$

Thus, we have

$$P_{\mu\nu}k_r^\mu U_r^\nu = k_{r\mu}U_r^\mu \approx k_{r0} + k_{r1}\Omega y_r - k_{r2}\Omega x_r, \tag{4.A6}$$

$$\eta_{\mu\nu}k'^{\mu}U'^{\nu} = k'_0,$$

for the rotating (Earth) frame F_r and the inertial frame F'.

The transformations of the covariant wave 4-vector between the inertial frame F' and a rotating frame F_r are given by the rotational Doppler effects,[29]

$$k'_0 \approx k_0 + \Omega y_r k_{r1} - \Omega x_r k_{r2}, \tag{4.A7}$$

$$k'_1 \approx k_{r1}\cos\Omega w_r - k_{r2}\sin\Omega w_r,$$

$$k'_2 \approx k_{r1}\sin\Omega w_r + k_{r2}\cos\Omega w_r,$$

where $k'_{\mu} = (k'_0, k'_1, 0, 0)$, $k'_{\mu}k'^{\mu} = 0$, $k'_1 = k'_0 > 0$ and $k'_2 = 0$. It follows from (4.A7) that

$$k_{r1} \approx \frac{-k_{r2}\cos\Omega w_r}{\sin\Omega w_r}, \qquad k_{r2} \approx -k'_0\sin\Omega w_r,$$

$$k_{r0} \approx k'_0[1 - \Omega y_r\cos\Omega w_r - \Omega x_r\sin\Omega w_r], \tag{4.A8}$$

where k'_0 is time-independent because the blackbody is at rest in the F' frame, while k_{r0} measured in the rotating frame F_r depends on the time w_r. Therefore, the general-frame covariant blackbody distribution (4.A2) in F_r and the LP invariant blackbody distribution (4.10) in F' frame are related by

$$B(P_{\mu\nu}k_r^{\mu}U_r^{\nu}, T) = \frac{1}{[exp(P_{\mu\nu}k_r^{\mu}U_r^{\nu}/k_BT) - 1]}$$

$$\approx \frac{1}{(exp(k_{r0}[1 - x_r\Omega\,\sin\Omega w_r - y_r\Omega\cos\Omega w_r]^{-1}/k_BT) - 1)} \tag{4.A9}$$

$$= B(k'_{\mu}U'^{\mu}, T) = \frac{1}{exp[k'_{\mu}U'^{\mu}/(k_BT)] - 1} = \frac{1}{exp[k'_0/(k_BT)] - 1},$$

where $T = T_{inv}$, and k'_0 is time-independent, and we have used (4.A7) and (4.A8) to simplify $P_{\mu\nu}k_r^{\mu}U_r^{\nu}$ in (4.A9).

Based on the principle of limiting continuation of physical laws in general reference frames, the rotational frequency shift (4.A8) implies that the general-frame (GF) covariant blackbody distribution $B(P_{\mu\nu}k_r^{\mu}U_r^{\nu}, T)$ measured in the Earth frame (revolving around the sun) is the same as the blackbody distribution $B(k'_{\mu}U'^{\mu}, T)$ in the inertial frame F'. Therefore, the GF covariant Planck distribution in (4.A9) does not depend on the time

w_r due to (4.A8), and hence, it does not have a 6-month variation. However, the GF non-covariant blackbody energy spectrum W_r, which is the quantity directly measured in satellite experiments, will display a 6-month variation, as shown in Eq. (4.9) in Sec 4.4.

Appendix 3 to Chapter 4.   Unphysical relativistic Planck law with a temperature 4-vector for CMB radiation

It is mathematically possible for the angle-dependent temperature (4.3) in Sec. 4-3 to be embedded in the Lorentz-Poincaré (LP) framework, provided the 'Landau-Lifshitz lemma',[24] is used. However, the results are not supported by experiments, as we shall demonstrate below.

Let us consider another inertial frame $F''(w'', x'', y'', z'')$ moving with a constant velocity v relative to the frame $F(w, x, y, z)$ and having the temperature

$$T''(\theta'') = T\sqrt{1 - v^2}/(1 + v \cos\theta''), \qquad (4.A10)$$

which has the same form as that in (4.3). Note that the angle θ' in (4.3) is defined through the wave 4-vector k'_μ, i.e.,

$$\cos\theta' = k'_x/k'_0. \qquad (4.A11)$$

Using the relativistic transformations for the angles $\cos\theta'$ and $\cos\theta''$ and the relativistic velocity-addition laws for V and v, i.e.,

$$\cos\theta' = (u + \cos\theta'')/(1 + u \cos\theta''), \qquad (4.A12)$$

$$u = \frac{v - V}{1 - vV}, \qquad (4.A13)$$

one can show that the temperatures $T''(\theta'')$ and $T'(\theta')$ are related by

$$T'' = T'\sqrt{1 - u^2}/(1 + u \cos\theta''), \qquad (4.A14)$$

which also has the desired form as (4.3). Because the angles are defined by $\cos\theta' = k'_x/k'_0$, all these consistent results are actually related to the Landau-Lifshitz lemma:[24, 31]

"The ratio $dp_x dp_y dp_z/E$ is an invariant quantity, since it is the ratio of corresponding components of two parallel four-vectors."

This property can be applied to the ratio k_0/T_0 of the fourth components of the wave 4-vector k_μ and the temperature 4-vector T_μ. In other words, there is a temperature 4-vector T'_μ that is proportional to the wave four-vector k'_μ in any inertial frame F'.

In this way, one has a well-defined angle in (4.A12). As a result, the ratio $k'_0/T' = k''_0/T''$ is indeed Lorentz invariant, i.e., a scalar.[g]

To demonstrate that the relations (4.A20) and (4.A24) are equivalent to assuming that the temperature 4-vector T_μ is proportional to the wave 4-vector k_μ, we write the transformations of T_μ and k_μ between two inertial frames F and F'',

$$T''_0 = \frac{T_0 - vT_x}{\sqrt{1 - v^2}}, \quad T''_x = \frac{T_x - vT_0}{\sqrt{1 - v^2}}, \quad T''_y = T_y, \quad T''_z = T_z; \qquad (4.A15)$$

$$k''_0 = \frac{k_0 - vk_x}{\sqrt{1 - v^2}}, \quad k''_x = \frac{k_x - vk_0}{\sqrt{1 - v^2}}, \quad k''_y = k_y, \quad k''_z = k_z; \qquad (4.A16)$$

$$T''_\mu = \alpha k''_\mu, \quad T_\mu = \alpha k_\mu, \quad T''_0 = T'', \quad T_0 = T, \qquad (4.A17)$$

where α is a proportional constant.

The inverse transformation of the temperature (4.A15) and the wave vector k_μ are

$$T_0 = \frac{T''_0 + vT''_x}{\sqrt{1 - v^2}} = \frac{T''_0(1 + v\,cos\theta'')}{\sqrt{1 - v^2}}, etc. \qquad (4.A18)$$

$$k_0 = \frac{k''_0 + vk''_x}{\sqrt{1 - v^2}} = \frac{k''_0(1 + v\,cos\theta'')}{\sqrt{1 - v^2}}, etc. \qquad (4.A19)$$

where the angle θ'' is defined by

$$cos\theta'' = \frac{k''_x}{k''_0} = \frac{T''_x}{T''_0}. \qquad (4.A20)$$

Thus, we have demonstrated that (4.A18) is the same as the angle-dependent temperature (A.4.10) with $T = T_0$. Since the F' frame moves with a velocity V relative to F, we have

$$k_0 = \frac{k'_0 + Vk'_x}{\sqrt{1 - V^2}}, \quad k_x = \frac{k'_x + Vk'_0}{\sqrt{1 - V^2}}, \quad etc. \qquad (4.A21)$$

---

[g] If one introduces the 4-dimensional momentum space, then $dp_x dp_y dp_z$ can be considered as the zeroth component of an element of the hypersurface determined by $p_\mu p^\mu = m^2$ of a particle. The element of a hypersurface of a 4-vector is directed along the normal to the hypersurface. In this case, the direction of the normal is the same as that of $p^\mu = (p_0, p_x, p_y, p_z)$. Thus, the ratio $dp_x dp_y dp_z/p_0$ is invariant because it is the ratio of corresponding components of two parallel 4-vectors, just like $k_0/T = k'_0/T'$.[24, 28]

which are consistent with (4.12) because $T_\mu = \alpha k_\mu$ and $\omega = \omega'[1 + V\cos\theta']/\sqrt{1-V^2}$ in (4.14) with $k'_0 = \omega'$, $k_0 = \omega$. From Eqs. (4.A16) and (4.A21), we obtain

$$\cos\theta'' = \frac{\cos\theta - v}{1 - v\,\cos\theta}, \qquad \cos\theta = \frac{\cos\theta' + V}{1 + V\,\cos\theta'}, \qquad (4.A22)$$

where we have used (4.A20), $\cos\theta = k_x/k_0 = T_x/T_0$, $\cos\theta' = k'_x/k'_0 = T'_x/T'_0$. Eliminating $\cos\theta$ in (4.A22), we can obtain (4.A12), as one would expect. The velocity-addition law (4.A13) follows from the coordinate transformations.

From (4.A21) and (4.3) in Sec. 4-3 with $T = T_0$, we obtain

$$\frac{k_0}{T_0} = \frac{k'_0}{T'_0}. \qquad (4.A23)$$

To summarize, from equations (4.A18), (4.A19) and (4.A23), we obtain the invariant relations,

$$\frac{k_0}{T} = \frac{k''_0}{T''} = \frac{k'_0}{T'}, \qquad (4.A24)$$

where we have used $T = T_0$, $T' = T'_0$, $T'' = T''_0$ and $T'_\mu = \alpha k'_\mu$. Since $k'_0 = \omega'$ and $k_0 = \omega$, the result (4.A24) implies that the Planck distributions (4.1) and (4.2) are invariant.

However, the result (4.A24) is obtained on the basis of the properties $k_\mu \propto T_\mu$, $k'_\mu \propto T'_\mu$ and $k''_\mu \propto T''_\mu$. Since $k_\mu k^\mu = 0$, etc., these properties imply that the temperature 4-vector is also a null vector in inertial frames F, F' and F'':

$$T_\mu T^\mu = 0, \qquad T'_\mu T'^\mu = 0, \qquad T''_\mu T''^\mu = 0. \qquad (4.A25)$$

However, in the past century, there has been no experimental support for (4.A25).[18-23]

Therefore, we conclude that the temperature 4-vector with the properties in (4.A25) and the Lorentz invariant Planck distribution based on (4.A24) are not supported either theoretically or experimentally.

References

1. T. D. Lee, *Particle Physics and Introduction to Field Theory* (Hardwood Academic, 1981) Ch. 14.
2. J. P. Hsu, L. Hsu and D. Katz, Mod. Phys. Letts. A **33**, 1850116 (2018); L. Hsu and J. P. Hsu, Experiments on the CMB Spectrum, Big Jets Model and Their Implications for the Missing Half of the Universe, DOI: https://doi.org/10.1051/ep jconf/201816801012.

3. F. J. Dyson, Bull. Amer. Math. Soc. **78** (1972).

4. P. A. Zyla et al. (Particle Data Group), Prog. Theor. Exp. Phys. 2020, 083C01 (2020).

5. A. Friedmann, Z. Phys. **10**, 377 (1922); G. Lemaître, Ann. Soc. Sci. de Bruxelles (in French), **A47**, 49 (1927); E. P. Hubble, Proc. Nat. Acad. Sci. **15**, 168 (1929); G. Gamow, Phys. Rev. **70**, 572(1946); R. A. Alpher, Phys. Rev. **74**, 1577 (1948).

6. R. A. Alpher, H. Bethe and G. Gamow, Phys. Rev. **73**, 803 (1948); R. A. Alpher, and R. C. Herman, Nature 162: 774 (1948).

7. A. A. Penzias and R. W. Wilson, Astrophys. J. Letters. 142: 419 (1965).

8. Ya. B. Zel'dovich, Advances Astron. Astrophys. **3** 242 (1965); H. Y. Chiu, Phys. Rev. Letter. **17**, 712, (1966); A. D. Dolgov, Ya. B. Zel'dovich, Reviews of Modern Physics, **53**, 1, January 1981.

9. Planck, ESA Science and Technology, 'HEMISPHERIC ASYMMETRY AND COLD SPOT IN THE COSMIC MICROWAVE BACKGROUND'. Google search.

10. T. Rothman and E. C. G. Sudarshan, *Doubt and Certainty* (Reading, Mass., Helix Book, 1998) pp. 214–264.

11. T. D. Lee and C. N. Yang, in *100 Years of Gravity and Accelerated Frames, The Deepest Insights of Einstein and Yang-Mills* (Eds. J. P. Hsu and D. Fine, World Scientific, 2005), p. 155. The baryon conservation law was first postulated by E. C. G. Stueckelberg, Helv. Phys. Acta, **11**, 299 (1938). T. D. Lee, *Particle Physics and Introduction to Field Theory*, (Hardwood Academic, 1981), Chs. 10 and 14, pp. 184–187, pp. 746–752.

12. K. Huang, *Quarks, Leptons and Gauge Fields* (World Scientific, 1982) p. 17, p. 62, pp. 127–147 and pp. 241–242.

13. S. Weinberg, *Gravitation and Cosmology, Principles and Applications of the General Theory of Relativity* (John Wiley and Sons, 1972) pp. 520–522. In the fundamental frame (in which the CMB is isotropic), the energy-momentum tensor of the photons is

$$dT^{\mu\nu} = 2p^\mu p^\nu h^{-1} [e^{h\nu/kT_{\gamma 0}-1}]^{-1} \sin\theta d\theta d\phi \nu d\nu.$$

In the earth frame, these photons have an energy-momentum tensor

$$dT'^{\mu\nu} = 2p'^\mu p'^\nu h^{-1} [e^{h\nu'/kT'_{\gamma 0}-1}]^{-1} \sin\theta' d\theta' d\phi' \nu' d\nu',$$

where $T'_{\gamma 0}$ is an angle-dependent temperature,

$$T'_{\gamma 0} \equiv (\nu' T_{\gamma 0})/\nu = [1 - v^2]^{-1/2} [1 - v\cos\theta] T_{\gamma 0}.$$

See also a Lorentz invariant discussion based on the Landau-Lifshitz condition for an angle-dependent temperature 4-vector for CMB radiations in Appendix 3 in this Ch. 4.

14. P. T. Landsberg and G. E. A. Matias, Physics A, **340**, 92 (2004); T. S. Biro and P. Ván, Europhys. Lett. **89** 30001 (2010); C. Farias, V. A. Pinto and P. S. Moya, Scientific Reports **7**, 17657 (2017); Z. C. Wu, Europhys. Lett. **88**, 20005 (2009) and Ref. 8.

15. D. Scott and G. F. Smoot, "27. COSMIC MICROWAVE BACKGROUND", Sec. 27.2.2, for The Review of Particle Physics, K. A. Olive et al. (Particle Data Group), Chinese Phys. C **38**, 09000, (2014); (http://pdg.lbl.gov). D. J. Fixsen et al., Astrophys. J. **473** 576 (1996).

16. S. Weinberg, *Cosmology* (Oxford, 2008) p. 130.

17. C. L. Bennett et al., Astrophys. J. Suppler. **148**, 1 (2003); V. F. Weisskopf, Am. Sci. **71**, 473 (1983).

18. A. Einstein, Jahrb. Rad. u. Elektr, **4**,411 (1907).

19. M. Planck, Ann. d. Phys. **26**, 792 (1908).

20. H. Ott, Zeitschr. d. Phys. **174**, 70 (1963).

21. H. Arzelies, Nuovo Cimento **35**, 792 (1965).

22. P. T. Landsberg, Nature **212**, 571 (1966).

23. D. Cubero, J. Casado-Pascual, J. Dundel, P. Talkner and P. Hanggi, Phys. Rev. Lett. **19**, 170601 (2007); N. G. van Kampen, Phys. Rev. **173**, 295 (1968).

24. L. Landau and E. Lifshitz, *The Classical Theory of Fields*, (tr. M. Hamermesh, Pergamon Press, 1971), p. 29. (In the 1951 edition, it just said 'two four-vectors' rather than 'two parallel four-vectors' in explaining the invariant quantity $dp_x dp_y dp_z/E$.)

25. G. N. Pellegrini and A. R. Swift, Am. J. Phys. **63**, 694 (1995) and references therein.

26. J. P. Hsu and L. Hsu, in the Proceedings of the Twelfth International Conference on Gravitation, Astrophysics and Cosmology, Moscow, 2015 (World Scientific, 2016) p. 3.

27. J. P. Hsu and L. Hsu, Phys. Letters, A **196**, 1 (1994). Such a formulation of space-time symmetry based solely on the first principle of relativity is crucial for generalization to non-inertial frames. See also Refs. 15 and 16.

28. J. P. Hsu and L. Hsu, *Space-Time, Yang-Mills Gravity, and Dynamics of Cosmic Expansion* (World Scientific, 2020) pp. 23–25, pp. 27–28, p. 30, pp. 39–42, pp. 50–51, pp. 61–77, pp. 266–269.

29. L. Hsu and J. P. Hsu, Eur. Phys. J. Plus, **128**, 74 (2013). For the rotational Doppler effects, see Eq. (40) in this paper.

30. V. Fock, *The Theory of Space Time and Gravitation* (tr. N. Kemmer, Pergamon Press, 1958) p. 151.

31. J. D. Jackson, *Classical Electrodynamics* (John Wiley & Sons, New York. 2nd ed.) pp. 223–225, pp. 608–611.

32. L. Landu and E. Lifshitz, *The Classical Theory of Fields* (tr. M. Hamermesh, Addison-Wesley, 1951) p. 31. After Eq. (2.19), the statement '...since it is the ratio of corresponding components of two four-vectors.' was replaced by '... since it is the ratio of corresponding components of two parallel four-vectors' in a later edition, (Pergamon Press, 1971) p. 29, after Eq. (10.1)

5

The Evolution of the Universe

5-1. Two models for the evolution of the universe

The expansion of the cosmos, cosmic redshifts, and the motion of galaxies have been investigated with two conceptually different models:

(i) The FLRW (Friedmann-Lemaire-Robertson-Walker)[1] model is based on general relativity (or Einstein gravity), which is based on the principle of general coordinate invariance in curved space-time. In this model, the universe begins with a primordial Big Bang. However, the absence of anti-matter in the observable portion of the universe remains a mystery.

(ii) The HHK (Hsu-Hsu-Katz)[1] model is based on Broader Particle-Cosmology with Yang-Mills gravity in flat space-time. In this model, the universe begins with a primordial Big Jets event, which evolves to form separate matter and anti-matter half-universes.

In both models, the gravitational field equations and the cosmological principle dictate the dynamics of galactic motion and cosmic evolution. Although the HHK model is formulated in flat space-time with inertial frames, quantum Yang-Mills gravity with translational gauge symmetry yields a Hamilton-Jacobi type equation, which we call the Einstein-Grossmann equation. This equation of motion involves an effective metric tensor $G_{\mu\nu}(x)$, which is derived in the geometric-optics limit of the wave equations for quantum particles. This effective metric tensor $G_{\mu\nu}(x)$ dictates the motion of classical objects and light rays and causes them to behave as if they were in a curved space-time, even though the actual underlying space-time of quantum particles and gravitational fields is flat. Furthermore, as we shall see, at the super-macroscopic scales using a new 'cosmological principle of isotropy,' energy is conserved in Lagrangian dynamics, and these dynamics then lead to the cosmic Okubo equation, whose solutions imply an expansion-contraction evolution of the universe.

One final important difference between the two models is that in the conventional FLRW model, space itself is interpreted as expanding so that it is possible for two objects to appear to have a relative velocity that is larger than the speed of light.[a] However, in the HHK model with flat

---

[a]Currently accepted estimates of the radius of the universe are roughly 46.5 billion light years, even though estimates of the age of the universe are much less than 46.5 billion years.

spacetime, there is nothing weird happening to space itself and no two objects ever have relative velocities that appear to be larger than the speed of light. A discussion of the full implications of this difference is beyond the scope of this book.

5-2.   Energy conservation at super-macroscopic scales

In physics, all established and observable phenomena at microscopic and macroscopic scales are consistent with the conservation of energy. However, within the usual framework involving the cosmological principles of homogeneity and isotropy, the Lagrange equations imply that the energy of objects at super-macroscopic scales, such as galaxies, is not conserved. The reason for this is that under the traditional assumptions that the cosmos is homogeneous and isotropic, the cosmic metric tensors are diagonal, as well as explicitly time-dependent. We have $G_{\mu\nu}(t) = \eta_{\mu\nu}U^2(t)$.[b,2] (To facilitate a comparison between the FLRW and HHK models, we shall use t for the time coordinate, expressed in units of seconds, rather than w, with the dimension of length.) The cosmic action S_c for the motion of a galaxy also involves the time-dependent cosmic metric $ds_c^2 = ds_t^2$ at super-macroscopic scales. Because the Lagrangian itself is time-dependent, conservation of energy is violated.

Because this difficulty is associated with the cosmological principles of homogeneity and isotropy, it is present in both the FLRW and HHK models. It is less obvious in the FLRW model however, because the FLRW model is formulated in Riemannian geometry, in which the coordinates are local and have no meaning by themselves, as stressed by S. S. Chern in his Lecture Notes of Differential Geometry.[3] In a lucid discussion of coordinates and momenta in general relativity, Wigner wrote: 'The basic premise of this theory [the general theory of relativity] is that coordinates are only auxiliary quantities which can be given arbitrary values for every event.'[3,5] Thus, the values of energy E and $dE/dt \neq 0$ cannot be measured in the FLRW model, because they do not have the usual operational meaning.

Given this difficulty with the usual metric, the question naturally arises, how can one preserve the principle of energy conservation in Lagrangian dynamics for galactic motion? Because the time dependence of the metric is due to the approximation that the universe is homogeneous, if one were to assume only the isotropy of the universe at super-macroscopic scales,

---

[b]This relation corresponds to the simplified relation with A(t) = B(t) = U(t) in Eq. (17.1) in Ch. 17 of Ref. 2.

then the metric tensor would be a function of r only and consequently, the energy defined by the Lagrangian will be conserved.

We now explore a new cosmic model based on Yang-Mills gravity with Lagrangian dynamics set in inertial frames, which assumes only the isotropy of the cosmos and thus preserves the conservation of energy in the motion of galaxies. This new simple 'Cosmic Isotropy Principle' states:

The effective metric tensor of the super-macroscopic space-time is a function of r only and takes the diagonal form, $G_{\mu\nu}(r) = (G_{00}(r), G_{11}(r), G_{22}(r), G_{33}(r))$.

The cosmic action S for a point-like galaxy with mass M is assumed to be

$$S = -Mc \int ds = \int L dt, \qquad L = -Mc\sqrt{G_{\mu\nu}(r)\dot{x}^\mu \dot{x}^\nu}, \qquad (5.1)$$

$$ds^2 = G_{\mu\nu}(r)dx^\mu dx^\nu, \quad G_{\mu\nu}(r) = (G_{00}(r), G_{11}(r), G_{22}(r), G_{33}(r)). \quad (5.2)$$

We discuss the details of deriving $G_{\mu\nu}(r)$ below.

5-3. Derivation of the r-dependent effective metric tensor $G_{\mu\nu}(r)$ in Yang-Mills gravity

Based on Yang-Mills gravity in inertial frames, the cosmic isotropy principle for the motion of galaxies implies that the effective metric tensor $G_{\mu\nu}$ is a function of $r = (x^2 + y^2 + z^2)^{1/2}$ only and takes the diagonal form,

$$G_{\mu\nu}(r) = \eta_{\mu\nu} R^2(r), \qquad r \geq 0, \qquad (5.3)$$

$$ds^2 = \eta_{\mu\nu} R^2(r) dx^\mu dx^\nu, \qquad \eta_{\mu\nu} = (1, -1, -1, -1), \quad \mu, \nu = 0, 1, 2, 3.$$

This cosmic metric tensor (5.3) dictates the motion of galaxies and, as we shall see, leads to a new model of the universe with both expansion and contraction (EC) phases. We call this the EC model. The EC model contrasts sharply with models based on a time-dependent $G_{\mu\nu}(t)$, which imply a universe that expands forever. (See Eq. (5.43) below.)

In such a simplified EC model, Yang-Mills gravity and the cosmic isotropy principle lead to the following relations,

$$G_{\mu\nu}(r) = \eta^{\alpha\beta} J_{\alpha\mu}(r) J_{\beta\nu}(r), \qquad and \qquad (5.4)$$

$$J_{\mu\nu}(r) = \eta_{\mu\nu} R(r) = (R(r), -R(r), -R(r), -R(r)). \qquad (5.5)$$

These 'effective metric tensors' $G_{\mu\nu}(r)$ and $J_{\mu\nu}(r)$ in (5.4) and (5.5) are governed by Yang-Mills gravity in inertial frames, and are derived from the space-time translation (T_4) gauge invariant gravitational field equations,

$$g^2 S_{\mu\nu} = \partial_\lambda \left(J_{\lambda'\rho'} C_{\rho\mu\nu} \eta^{\rho\rho'} - J_{\lambda'\alpha'} C_{\alpha\beta}{}^\beta \eta^{\alpha\alpha'} \eta_{\mu\nu} + C_{\mu\beta}{}^\beta J_{\nu\lambda'} \right) \eta^{\lambda\lambda'} \quad (5.6)$$

$$-C_{\mu\alpha}{}^\beta \partial_\nu J_{\alpha'\beta} \eta^{\alpha\alpha'} + C_{\mu\beta}{}^\beta \partial_\nu J_{\alpha\alpha'} \eta^{\alpha\alpha'} - C_{\lambda\beta}{}^\beta \partial_\nu J_{\mu\lambda'} \eta^{\lambda\lambda'},$$

$$C_{\mu\nu\alpha} = J_{\mu\lambda}(\partial_{\lambda'} J_{\nu\alpha}) \eta^{\lambda\lambda'} - J_{\nu\lambda}(\partial_{\lambda'} J_{\mu\alpha}) \eta^{\lambda\lambda'}, \quad C_{\mu\alpha}{}^\beta = C_{\mu\alpha\beta'} \eta^{\beta\beta'} \quad (5.7)$$

where $g^2 = 8\pi G/c^3$ and μ and ν should be made symmetric in the field Eq. (5.6).[2,4] The quantity $S_{\mu\nu} = (1/2)[\bar\psi i \gamma_\mu \Delta_\nu \psi - i(\Delta_\nu \bar\psi)\gamma_\mu \psi]$ is the fermion source of the tensor fields $\phi_{\mu\nu}$ and $C_{\mu\nu\alpha}$ is the T_4 gauge curvature.[c]

Based on (5.5) and (5.7), we first calculate the useful components of the gauge curvature $C_{\alpha\beta\gamma}$ with the Minkowski metric $\eta_{\mu\nu} = (1, -1, -1, -1)$,

$$C_{1\beta\gamma} = R\partial_1 R \eta_{\beta\gamma} - R\partial_\beta R \eta_{1\gamma}, \quad \beta, \gamma = 0, 1, 2, 3; \quad (5.8)$$

$$C_{\alpha\beta}{}^\beta = 3R\partial_\alpha R, \quad C_{\rho 00} = R\partial_\rho R, \quad C_{\rho 11} = -R\partial_\rho R - R\partial_1 R \eta_{\rho 1},$$

$$C_{\rho 22} = -R\partial_\rho R - R\partial_2 R \eta_{\rho 2}, \quad C_{\rho 33} = -R\partial_\rho R - R\partial_3 R \eta_{\rho 3}.$$

We then use spherical coordinates to express (5.6) and (5.8) in order to find the spherically symmetric solution $R(r)$ in (5.5). For example, we have the following relations in terms of r,

$$\partial^\alpha C_{\alpha\beta}{}^\beta = -3(\partial_r R)^2 - (6R/r)\partial_r R - 3R(\partial_r)^2 R, \quad \partial_r = \partial/\partial r,$$

$$\partial_\alpha \partial^\alpha R(r) = -\bigtriangledown^2 R(r) = -(2/r)\partial_r R - (\partial_r)^2 R. \quad (5.9)$$

Just as in the FLRW and HHK models,[1,8] we assume that the fermion source $S_{\mu\nu}$ in (5.6) in Yang-Mills gravity is approximated by the effective energy-momentum tensor $T_{\mu\nu}$ of 'super-macroscopic matter':

$$T_{\mu\nu} = \left(\rho(r)R^2(r), -P(r)R^2(r), -P(r)R^2(r), -P(r)R^2(r) \right), \quad (5.10)$$

$$\rho(r) = \rho_o/R^3(r), \quad and \quad P(r) = \rho(r)\omega,$$

where $\rho(r)$ and $P(r)$ are the energy density and pressure of super-macroscopic bodies, respectively, for a matter-dominated cosmos. We also assume a simplified equation of state $P(r) = \rho(r)\omega$ in (5.10) for super-macroscopic matter.[1] It is very unlikely that we can determine the real

---

[c]As usual, $[\Delta_\mu, \Delta_\nu] = C_{\mu\nu\alpha}\partial^\alpha$, where $\Delta_\nu = J_{\nu\lambda}\partial^\lambda$ is the T_4 gauge covariant derivative, and $J_{\mu\nu} = \eta_{\mu\nu} + g\phi_{\mu\nu}$ depends the gravitational tensor field $\phi_{\mu\nu} = \phi_{\nu\mu}$.[2]

physical sources of these super-macroscopic properties at the present time. (One may speculate that their sources include effectively everything in the universe.)

After some tedious but straightforward calculations, (5.6)–(5.10) lead the following equations in Yang-Mills gravity with the cosmic isotropy principle:

$$g^2 T_{00} = g^2 \rho_o R^{-1} = 4R(\partial_r R)^2 + (4R^2/r)\partial_r R + 2R^2(\partial_r)^2 R, \qquad (5.11)$$

$$g^2 T_{11} = -4R(\partial_r R)^2 - (4R^2/r)\partial_r R - 2R^2(\partial_r)^2 R + 10R(\partial_1 R)^2 + 2R^2(\partial_1)^2 R,$$

$$g^2 T_{22} = -4R(\partial_r R)^2 - (4R^2/r)\partial_r R - 2R^2(\partial_r)^2 R + 10R(\partial_2 R)^2 + 2R^2(\partial_2)^2 R,$$

$$g^2 T_{33} = -4R(\partial_r R)^2 - (4R^2/r)\partial_r R - 2R^2(\partial_r)^2 R + 10R(\partial_3 R)^2 + 2R^2(\partial_3)^2 R.$$

We have seen that expressions for $g^2 T_{22}$ and $g^2 T_{33}$ are the same as that for $g^2 T_{11}$ with ∂_1 in the last two terms replaced by ∂_2 and ∂_3 respectively.

We are interested in the spherically symmetric solution $R(r)$. The Yang-Mills gravitational field Eq. (5.6), together with (5.4)–(5.5) and (5.7)–(5.10), leads to two independent equations. One is Eq. (5.11) for T_{00}. The other is the combination $T_{11} + T_{22} + T_{33}$,

$$g^2(T_{11} + T_{22} + T_{33}) = -2R(\partial_r R)^2 - (8R^2/r)\partial_r R - 4R^2(\partial_r)^2 R. \quad (5.12)$$

Both (5.11) and (5.12) involve the variable r only. Using the equations in (5.10)–(5.12) and assuming, as usual, the form $R(r) = b_o r^n$ with constant b_o and n to be determined, we obtain the following solutions for $R(r)$ in the EC model,

$$R(r) = b_o r^{1/2}, \qquad b_o^4 = \frac{2g^2 \rho_o c}{5}, \qquad \omega = \frac{7}{15}. \qquad (5.13)$$

With the help of the cosmic isotropy principle, Yang-Mills gravity exercises the full power of its T_4 gauge symmetry to determine a new scale factor $R(r)$ in (5.3)–(5.5). We shall call $R(r)$ the Sudarshan scale factor to distinguish it from the usual time-dependent Robertson-Walker scale factor in conventional cosmic models. Thus, Yang-Mills gravity with the cosmic isotropy principle dictates the dynamics of the motion of galaxies with the r-dependent Sudarshan scale factor $R(r) = b_o r^{1/2}$ in (5.13), which is intimately related to the conservation of energy. The Sudarshan scale factor $R(r)$ in the Okubo equation (see (5.17) below) will lead to the expansion-contraction (EC) model to be discussed later.

5-4. Super-macroscopic Lagrangian dynamics with energy conservation

The motion of an idealized point-like galaxy with a mass M is, as usual, assumed to be determined by the least action S_{EC} involving the effective metric tensor $G_{\mu\nu}(r) = \eta_{\mu\nu}R^2(r)$ in ds^2 or in the Lagrangian L_{EC},

$$S_{EC} = -Mc \int ds = -Mc \int \sqrt{\eta_{\mu\nu}R^2(r)dx^\mu dx^\nu} = \int L_{EC}dt, \qquad (5.14)$$

$$L_{EC} = -Mc^2 R(r)\sqrt{1 - \dot{r}^2/c^2}, \qquad R(r) = b_o r^{1/2},$$

where we have used spherical coordinates $x^\mu = (ct, r, \theta, \phi)$ with a constant θ and ϕ. According to L_{EC}, the trajectory of a distant galaxy is a radial line with constant θ and ϕ, consistent with the cosmic isotropy principle.

The EC Lagrangian (5.14) for a galaxy gives the following equation of motion,

$$\frac{d}{dt}\frac{\partial L}{\partial \dot{r}} - \frac{\partial L}{\partial r} = 0, \qquad r, t \geq 0, \qquad L \equiv L_{EC}, \qquad (5.15)$$

i.e.,
$$\frac{d}{dt}\frac{MR\dot{r}}{\sqrt{1 - \dot{r}^2/c^2}} + \frac{Mc^2 b_o}{2r^{1/2}}\sqrt{1 - \frac{\dot{r}^2}{c^2}} = 0.$$

Since L does not involve time explicitly, (5.15) leads to a constant energy E,

$$dE/dt = 0, \qquad E = \frac{\partial L}{\partial \dot{r}}\dot{r} - L = \frac{Mc^2 R}{\sqrt{1 - \dot{r}^2/c^2}}, \qquad (5.16)$$

where the conserved energy E of a galaxy involves the Sudarshan scale factor $R(r)$ and the recession velocity $\dot{r}$.

It is interesting to note that for galactic motion, the conserved cosmic energy E in (5.16) of a galaxy takes the form $E = (M_{eff})c^2/\sqrt{1 - \dot{r}^2/c^2}$, where $M_{eff} = MR(r)$ plays the role of an effective mass of a galaxy. In this sense, the Sudarshan scale factor $R(r)$ in $ds^2 = \eta_{\mu\nu}R^2(r)dx^\mu dx^\nu$ is a characteristic of the whole universe and dictates the motion of galaxies at the super-macroscopic scale, as shown in (5.15) and (5.16). It is intriguing that the effective mass M_{eff} of a galaxy appears to resemble Mach's idea that the inertia of a body is due to an (instantaneous) interaction with all masses of the universe. Unexpectedly, this property of an effective mass of a galaxy turns out to be realized in the galactic motion in the super-macroscopic world through the Lagrangian dynamics (5.14) involving the Sudarshan scale factor $R(r)$.

Based on the action S_{EC} in (5.14), we can also derive the cosmic Okubo equation of motion for distant galaxies (in the radial direction) in the form of the Hamilton-Jacobi type equation,

$$G^{\mu\nu}(r)(\partial_\mu S)(\partial_\nu S) - M^2 c^2 = R^{-2}[(\partial_{ct} S)^2 - (\partial_r S)^2] - M^2 c^2 = 0, \quad (5.17)$$

$$S = S_{EC}, \quad G^{\mu\nu} = \eta^{\mu\nu} R^{-2}(r), \quad R(r) = b_o r^{1/2}$$

which is equivalent to the Lagrangian Eq. (5.15). To find the solution of the Okubo Eq. (5.17), one writes, as usual, $S = -Et + f(r)$ because the energy E is conserved. The Okubo Eq. (5.17) gives $f(r) = \pm(1/c) \int \sqrt{E^2 - M^2 c^4 R^2}\, dr$.[2] As usual, the equation of motion is given by $\partial S/\partial E = constant$, which leads to $-t + \partial f(r)/\partial E = constant$. Differentiating with respect to the radius r and using $dt/dr = 1/\dot{r}$, the cosmic Okubo equation leads to the radial velocity of a galaxy,

$$\dot{r} = \pm c\sqrt{1 - r/R_o}, \quad R_o = E^2/(M^2 c^4 b_o^2). \quad (5.18)$$

The $\pm$ solutions for the radial motion in (5.18) indicate that both expanding and contracting solutions for the universe exist. The length R_o characterizes the size of the cosmos. These positive and negative radial velocities (5.18) can also be obtained from the Lagrangian Eq. (5.15).

Since we are dealing with an extremely simplified model of cosmic motion, it is natural to assume that the cosmic isotropy principle and hence, the EC Eq. (5.15) hold for all times $t \geq 0$, as well as for all velocities $0 < |\dot{r}| < c$ (where $|\dot{r}| \to c$, as $r \to 0$), as one can see from (5.18). In the EC model, it is not necessary to make ad hoc assumptions concerning whether cosmic evolution is dominated by relativistic or non-relativistic matter since both are automatically covered by the solution of the cosmic Okubo equation of motion for galaxies. As a result, the model gives the cosmic equation of motion for a galaxy during all epochs from the beginning to the end of cosmic history.

In the remainder of the chapter, we explore some further implications of this model, in which only the isotropy, but not the homogeneity, of the universe is assumed and the universe both expands and contracts. The model is obviously incomplete, as it does not incorporate all known interactions, including those that would cause an accelerated expansion.[d] However, it can be useful to explore the possible consequences of such a model, including where it might overlap with or reinterpret aspects of the current most widely accepted model, such as Hubble's law and cosmic redshifts.

---

[d]It appears natural and simple for the tensor gauge field $\phi_{\mu\nu}$ of gravity to contribute to the effective metric tensor $G_{\mu\nu} = \eta^{\alpha\beta}(\eta_{\alpha\mu} + g\phi_{\alpha\mu})(\eta_{\beta\nu} + g\phi_{\beta\nu})$ in (5.4) and (2.10). However, it appears to be not simple for a vector gauge field in a gauge invariant Lagrangian to contribute to the effective metric tensor $G_{\mu\nu}$.

5-5. An expanding and contracting universe due to the new cosmic isotropy principle

The trajectory of a galaxy can be determined by the equation of constant cosmic energy E in (5.16) or by the cosmic Okubo Eq. (5.17). The expanding and contracting (EC) model for the cosmos gives the exact expansion and contraction velocities $\dot{r}$ of a galaxy as a function of r, as measured in an inertial frame,

$$Expansion: \qquad \dot{r} = c\sqrt{1 - r/R_o} ; \qquad (5.19)$$

$$Contraction: \qquad \dot{r} = -c\sqrt{1 - r/R_o} ,$$

where R_o is given by (5.18). The results (5.19) of the EC model predict that the universe is finite because $r < R_o = E^2/(M^2 c^4 b_o^2)$ in both expansion and contraction parts of the cycle. Although R_o is different for each galaxy (depending on the mass M and energy E of any particular galaxy), they are all confined to a finite region in space. Although the finite size of the region of space occupied by galaxies suggests the possibility of the existence of multiple cosmos in a limitless space-time, we shall not speculate about such a possibility.

The constant R_o in (5.18) can be roughly estimated. Using the relationship between b_o^4 and ρ_o in (5.13) and assuming the present critical density[8] $\rho_{o,crit} = (3H_0^2)/(8\pi G)$ for ρ_o, we estimate that

$$R_o = \frac{E^2}{M^2 c^4 b_o^2} > 10^{26} m, \qquad E/(Mc^2) > 1, \qquad (5.20)$$

where we have used $g^2 c^3 = 8\pi G$,[2,3] and the Hubble time $1/H_0 \approx 3 \times 10^{17} sec$.[8] This result (5.20) from the EC model is consistent with observations. Although the constant cosmic energy E and the mass M of a galaxy are not known, it is natural to assume that $E/Mc^2 \geq 1$ for moving galaxies for purposes of estimating R_o.

The motion of a galaxy is complicated because the recession velocity (5.19) during the expansion part of the cycle implies that it has a non-zero negative acceleration $\ddot{r}$,

$$\ddot{r} = \frac{-c^2}{2R_o} \quad < 0. \qquad (5.21)$$

Roughly speaking, since $1/R_o$ or c/R_o is extremely small (because R_o is extremely large, as estimated in (5.20)), the deceleration of a galaxy in (5.21) could be roughly estimated using (5.20). We obtain

$$|\ddot{r}| < 4.5 \times 10^{-10} m/sec^2. \qquad (5.22)$$

Such a small deceleration in (5.22) can simplify our discussion of cosmic redshifts with light sources at rest relative to non-inertial galaxies. In order to understand the observed late-time cosmic acceleration, the effect of the long range Okubo force due to conserved baryonic charges should be considered.[2,9] However, such an inclusion could be quite complicated because, for example, the boundary conditions for physical fields in the EC model with a finite R_o may be different from those of the conventional field theory.

We can integrate (5.19) to obtain a relation between the distance r and the time t,

$$r = R_o - \frac{c^2}{4R_o}(f - t)^2, \tag{5.23}$$

where f is a constant of integration. In order to determine f in the expansion part of the cycle, we define the following initial condition: the position of the galaxy is $r(0) > 0$ at time $t = 0$, thus we have

$$f = \frac{2}{c}\sqrt{R_o(R_o - r(0))}. \tag{5.24}$$

5-6. Cosmic expansion with an approximate Hubble's linear law

To see whether the expansion part of the EC model is consistent with Hubble's linear law,[6,12] we first express the recession velocity as a function of time involving the 'Hubble function' $H = H(t) = 1/t$. When one sets $t = t_o \approx 10^{10}$ years to be the present age of the universe, $H(t_o)$ is the Hubble constant, which can be determined experimentally. From (5.23), (5.24) and (5.19), we obtain

$$t = \frac{2R_o}{c}\left(\sqrt{1 - \frac{r(0)}{R_o}} - \sqrt{1 - \frac{r}{R_o}}\right), \quad 0 < t < \frac{2R_o}{c}\sqrt{1 - \frac{r(0)}{R_o}}, \tag{5.25}$$

where (5.25) implies that the time interval of the expansion cycle is from $t = 0$ to a maximum time $t = (2R_o/c)\sqrt{1 - r(0)/R_o} \equiv T_o$. Using the definition $H(t)t = 1$, we obtain the usual form of Hubble's law

$$\dot{r} = c\sqrt{1 - \frac{r}{R_o}} = H(t)tc\sqrt{1 - \frac{r}{R_o}} \tag{5.26}$$

$$= H(t)\left[2r - 2R_o + 2R_o\sqrt{\left(1 - \frac{r}{R_o}\right)\left(1 - \frac{r(0)}{R_o}\right)}\,\right],$$

$$\approx H(t)\left[r - \frac{r^2}{4R_o}\right] > 0, \qquad r(0) << r << R_o. \tag{5.27}$$

Equation (5.26) or (5.19) is the exact Hubble's law for all time t predicted by the EC model. The simplified result (5.27) is obtained in the approximations, $r/R_o << 1$ and $r(0)/r << 1$. It involves both a linear term and a quadratic term. The quadratic term in recession-velocity-distance was discussed by Segal et al.[10] We have the usual Hubble's linear law,

$$\dot{r} \approx H(t)r, \qquad t = t_o, \tag{5.28}$$

only when r/R_o is negligible, which could be interpreted that the recession velocity is approximately proportional to the distance r, provided that, at the present age of the universe $t = t_o \approx 10^{10}$ years, $H(t)$ can be approximated by the Hubble constant $H(t_o)$ in analyzing the velocity-distance relation of galaxies. Thus, for the present state of the universe, approximation (5.27) is consistent with the expansion part of the cycle in the EC model.

The exact relation between the recession velocity $\dot{r}$ and distance r for galaxies for all time is given by (5.26), i.e.,

$$\dot{r} = c\sqrt{1 - \frac{r}{R_o}}.$$

In the future, when we have data that covers a wide enough range of time such that $H(t)$ in (5.28) cannot be approximated as a constant $H(t_o)$, then one can test the difference between the exact relation (5.26) for the recession velocity and the approximation (5.28).

The EC model with (5.25) leads to the following expansion picture of the universe for $0 < t < T_o = (2R_o/c)\sqrt{1 - r(0)/R_o}$:

$$r(t) = R_o\left[1 - \left(\sqrt{1 - \frac{r(0)}{R_o}} - \frac{ct}{2R_o}\right)^2\right] \tag{5.29}$$

$$= R_o - \frac{c^2(T_o - t)^2}{4R_o}, \qquad T_o = \frac{2R_o}{c}\sqrt{1 - \frac{r(0)}{R_o}},$$

$$\dot{r}(t) = c\left(\sqrt{1 - \frac{r(0)}{R_o}} - \frac{ct}{2R_o}\right), \qquad \ddot{r} = \frac{-c^2}{2R_o} < 0. \tag{5.30}$$

At the initial time $t = 0$ and final time $t = T_o$ of the expansion part of the cycle, the position, velocity and constant acceleration of a galaxy are given by (5.29) and (5.30),

$$\text{initial time } t = 0: \quad r = r(0) > 0, \quad \dot{r}(0) = c\sqrt{1 - \frac{r(0)}{R_o}} \approx c, \qquad (5.31)$$

$$\text{final time } t = T_o: \quad r(T_o) = R_o, \qquad \dot{r}(T_o) = 0,$$

where the initial radius $r(0)$ is negligible in comparison with the maximum radius R_o of the cosmos.

Thus, the conservation law of the cosmic energy (5.16) (or, equivalently the Lagrange Eq. (5.15) for the motion of galaxies) in the EC model leads to the solution (5.31), which implies a 'detonation' at the beginning of the expansion. In other words, the initial velocity is almost the speed of light, $\dot{r}(0) = c\sqrt{1 - r(0)/R_o} \approx c$ because the initial radius $r(0)$ is expected to be much smaller than the maximum radius R_o of the cosmos.

5-7. Death and re-birth of the universe due to energy conservation

The time t of the contraction part of the cycle is given by

$$T_o \leq t \leq 2T_o. \qquad (5.32)$$

This can be verified as follows: The contraction velocity is negative along the radial direction, as given in (5.19),

$$\dot{r}(t) = -c\sqrt{1 - r(t)/R_o} < 0. \qquad (5.33)$$

We can integrate this equation to obtain $r - R_o = -(c^2/4R_o)(t - f'/c)^2$. The constant of integration f' is defined so that the position of a galaxy is $r = R_o$ at time $t = T_o$, which gives $f' = cT_o$. Thus, we obtain

$$r(t) = R_o - \frac{c^2}{4R_o}(t - T_o)^2, \quad r(T_o) = R, \quad r(2T_o) = r(0), \qquad (5.34)$$

$$2T_o \geq t \geq T_o = \frac{2R_o}{c}\sqrt{1 - \frac{r(0)}{R_o}}.$$

One can verify that at time $t = 2T_o$, (5.34) gives $r(2T_o) = r(0)$, i.e., the galaxy completes its full cycle of expansion-contraction and returns to its original position $r(0)$ at time $t = 0$.

We stress that the exact relation for the velocity $\dot{r}(t)$ and radial distance $r(t)$ is given by (5.33) for the time interval defined in (5.32) for the contraction part.

One may wonder whether this EC model in the contraction part is consistent with the usual Hubble's linear law. We demonstrate that regardless of whether one uses the Hubble function in the form $H(t)t = 1$ with t defined in the relation (5.32) or uses $H(t')t' = 1$ with $t' = t - T_o$, the resulting relation between the velocity $\dot{r}(t)$ and the distance r is not the same as the usual Hubble's law, in contrast to (5.28) for the expansion cycle.

To wit, suppose one expresses the 'approach velocity' (as contrasted with the 'recession velocity') as a function of time involving the 'Hubble function' $H(t') = 1/t'$, where $t' = t - T_o$ for the contraction part of the cycle. From (5.19) for the contraction part of the cycle and (5.34), we obtain

$$t - T_o = \frac{2R_o}{c}\sqrt{1 - \frac{r}{R_o}}, \qquad 2T_o \geq t \geq T_o. \qquad (5.35)$$

The radial velocity of a galaxy is negative in the contraction part of the cycle. It follows from (5.33), (5.35) and $H(t')t' = 1$ that

$$\dot{r}(t) = -H(t')t'c\sqrt{1 - \frac{r}{R_o}} = H(t')(-2R_o)\left[1 - \frac{r}{R_o}\right] \qquad (5.36)$$

$$= -H(t')2\left[R_o - r(t)\right] < 0, \qquad T_o \geq t' = t - T_o \geq 0. \qquad (5.37)$$

Or equivalently, suppose we define $H(t)t = 1$, we have

$$\dot{r}(t) = -H(t)tc\sqrt{1 - \frac{r}{R_o}} = -H(t)\left[cT_o + 2R_o\sqrt{1 - \frac{r}{R_o}}\right]\sqrt{1 - \frac{r}{R_o}}. \qquad (5.38)$$

The result (5.37) or (5.38) is the new law for approach velocities (measured in an inertial frame) for a galaxy in the contraction part of the cycle.

The EC model gives the following contraction picture of the universe for all times $2T_o \geq t \geq T_o$:

$$r(t) = R_o - \frac{c^2}{4R_o}(t - T_o)^2, \qquad T_o < t < 2T_o, \qquad (5.39)$$

$$\dot{r}(t) = \frac{-c^2}{2R_o}(t - T_o) < 0, \qquad \ddot{r} = \frac{-c^2}{2R_o} < 0.$$

At time $t = T_o$ and time $t = 2T_o$ of the contraction part of the cycle, we have

$$r(T_o) = R_o, \qquad \dot{r}(T_o) = 0, \qquad (5.40)$$

$$r(2T_o) = r(0), \qquad \dot{r}(2T_o) = -c\sqrt{1 - r(0)/R_o} \approx -c,$$

where $r(0) << R_o$. Thus, the EC model gives a picture of the contraction part of the cycle, which begins with a galaxy at rest at time $t = T_o$ and $r = R_o$. The final negative velocity reaches $\approx -c$ at time $t = 2T_o$ at the position $r(0)$ due to the constant negative acceleration $\ddot{r} = -c^2/(2R_o)$ in (5.39).

When one compares the initial velocity $\dot{r}(0) = c\sqrt{1 - r(0)/R_o} \approx c$ in (5.31) and the final velocity $\dot{r}(2T_o) = -c\sqrt{1 - r(0)/R_o} \approx -c$ in (5.40), the EC model based on the Broader Particle-Cosmology pictures the contraction part of cosmic evolution as an enormous number of particles from all directions moving toward the initial region at nearly the speed of light, similar to an extremely high energy particle collider. (Each galaxy moves back to its initial position $r(0)$.) Assuming that all known fundamental laws of physics remain intact throughout the cosmic evolution, it appears that there is only one possible physical result: namely, the creation of a new 'baby universe' and a 'detonation' that starts a new expansion. With the exception of energies released in the form of electromagnetic and gravitational radiations (negligible compared to the energy of galaxies), etc., all of the initial energy of the cosmos will still be present at the end of the contraction phase.

When one considers the motion of a photon with zero mass, the Lagrangian in (5.14), its Eq. (5.15) and the resultant energy in (5.16) are not well-defined for the limit $M \to 0$. Nevertheless, one can use the Okubo equation

$$G^{\mu\nu}(r)\partial_\mu S \partial_\nu S - M^2 = 0 \qquad (5.41)$$

with $M = 0$. This equation can be used as the classical law of motion for a massless particles or photon. One has the trajectory $\dot{r} = c$ for the massless photon. It is similar to (5.19) with $R_o = E^2/(M^2 c^4 b_o^2) \to \infty$, so that $\dot{r} = c$ for all times. The maximum time $T_o = (2R_o/c)\sqrt{1 - r(0)/R_o}$ for the expansion approaches ∞ as $M \to 0$. As a result, there is no contraction cycle for massless particles in the model. Thus, in sharp contrast to the massive galaxies, the EC model suggests that massless particles will leave the region of the cosmos occupied by massive galaxies and escape to infinity, never to return.

5-8. Cosmic redshifts in the Expansion-Contraction model

The cosmic redshift z of light emitted from a distant galaxy involves a small constant acceleration, as shown in (5.22). This redshift can be derived as an accelerated Wu-Doppler effect with the help of the accelerated Wu transformations,[2, 11] which reduce to the Lorentz transformations in the limit of zero acceleration. The wave 4-vector $K_{e\mu}$ emitted from a distant galaxy is associated with an eikonal equation and $\partial_\mu S \to \partial_\mu \psi_e = k_{e\mu}$. The wave 4-vector of light as measured in an inertial frame satisfies the usual eikonal equation $\eta^{\mu\nu}\partial_\mu\psi\partial_\nu\psi = 0$ with $\partial_\mu\psi = k_\mu$. It is natural to treat the observed cosmic redshift based on the 'covariant eikonal equation,' $G^{\mu\nu}(r)\partial_\mu\psi_e\partial_\nu\psi_e = \eta^{\mu\nu}\partial_\mu\psi\partial_\nu\psi$, which satisfies the principle of limiting continuation of physical laws.[2] With the help of the weak equivalence principle, the cosmic redshift z of a non-inertial galaxy is[7]

$$z = \frac{1 + \dot{r}/c}{\sqrt{1 - \dot{r}^2/c^2}} - 1 \approx \frac{\dot{r}}{c} + \frac{\dot{r}^2}{2c^2}, \tag{5.42}$$

$$or \qquad \frac{\dot{r}}{c} = \frac{2z + z^2}{2 + 2z + z^2}, \tag{5.43}$$

as measured in an inertial frame. According to the EC model, the cosmic redshift $z = 0$ when $\dot{r} = 0$; and $z \to \infty$ when $\dot{r}$ approaches the speed of light c. We note that $\dot{r}$ in (5.43) is the non-constant recession velocity of a distant galaxy at the moment when the light was emitted.

The usual FLRW model also formally has properties similar to those in (5.19) and (5.28) in the EC model. Suppose one considers (5.42) with $a(t) = a_0 t^{2/3}$ in Lagrangian dynamics with the action $S_F = \int(-Mc)ds_F$ involving the metric tensor $g_{\mu\nu}(t)$. Following the steps corresponding to (5.14)–(5.19) and (5.26)–(5.28), one obtains

$$\dot{r}(t) \approx H(t)\frac{r(t)}{2}, \qquad \dot{r}(t) = \frac{9}{a_o^3}\frac{1}{r^2(t)}, \tag{5.44}$$

$$H(t) = \frac{\dot{a}}{a} = \frac{2}{3t}, \qquad r(t) = \frac{3t^{1/3}}{a_o},$$

where the 'relativistic condition' ($mc \ll p$) is used for simplicity.[2] Note that the proportional relation $\dot{r}(t) \propto 1/r^2(t)$ holds for all times and is derived from Lagrangian dynamics in the conventional FLRW model. The usual interpretation of $\dot{r}(t) \propto H(t)r(t)/2$ in (5.44) is that the velocity $\dot{r}(t)$ of a distant galaxy is proportional to its distance $r(t)$ for only a certain limited time interval during the evolution of the universe, rather than for all times.

5-9.   The Expansion-Contraction (EC) model versus a Permanent Expansion model

Quantum Yang-Mills gravity in flat space-time, together with the cosmological principle of homogeneity and isotropy, leads to the Okubo equation $G^{\mu\nu}(t)\partial_\mu S\partial_\nu S - m^2 = 0$ for the motion of a galaxy and for cosmic redshifts. The solutions to Okubo equations sketch a total cosmic history as follows:[7]

I. In the beginning, at $t = 0$, the universe has an initial radius $r_o > 0$. It expands with a velocity $\dot{r} = C_o$, and the initial redshift is $z = \infty$. (The universe begins with a 'detonation.')

II. At the end, $t \to \infty$, the final velocity $\dot{r} \to 0$, the final radius $r \to \infty$, and the final redshift $z \to 0$. (The universe ends without a 'sound.')

All these properties are intertwined with the non-conservation of energy in the motion of a galaxy. Or more to the point, all this is the logical result of the conservation of the radial momentum in the Okubo equation $G^{\mu\nu}(t)\partial_\mu S\partial_\nu S - m^2 = 0$ involving the time-dependent effective metric tensor $G^{\mu\nu}(t)$, which is independent of r.

To resolve the problem of energy non-conservation, we formulate a dynamics of the cosmos on the basis of Yang-Mills gravity together with a new cosmic isotropy principle, so that the resultant effective metric tensor $G^{\mu\nu}(r)$ and the Sudarshan scale factor $R(r)$ depend on the radial position r, but not the time t. Consequently, energy is conserved in galactic motion. This cosmic dynamics leads to a cosmos that both expands and contracts with a radial velocity $\dot{r} = \pm c\sqrt{1 - r/R_o}$.

Two interesting predictions of the EC model are:

(A) Based on the critical present density $\rho_{o,crit} = (3H_0^2)/(8\pi G)$, the EC model predicts two new important features of the universe, i.e., the maximum finite size R_o in (5.20) of the universe and the time T_o to complete the expansion part of the cycle,

$$R_o > 10^{10} \; light \; years, \qquad T_o = \frac{2R_o}{c} > 2.2 \times 10^{10} \; years, \qquad (5.45)$$

both of which are consistent with current cosmological data.[8]

(B) During the contraction part of the cycle, all particles in the universe accelerate back to the initial spatial region from which they came, attaining velocities close to the speed of light at the end. Such a final state will lead to a gigantic explosion due to the strong interactions[13–16] of the enormous number of extremely high-energy particles in a 'cosmic super-collider.' According to the fundamental CPT invariance[13] in Broader Particle-Cosmology, all interactions of particles must satisfy a maximum symmetry between particles and anti-particles. The experimental observation that there are no equal amount of antiparticles in our observable portion of the universe suggests that the universe could have been created as two big jets rather than one big bang.[1] It is natural to apply Lagrangian dynamics to each jet, whose center of mass is approximately at rest in an inertial frame. Broader Particle-Cosmology with particle physics and quantum fields theory, together with the conservation of energy, provides a reliable way to obtain some understanding of such a spectacular re-birth of the universe every $T_o > 2.2 \times 10^{10}\ years$.

In Table 4.1, we summarize and compare physical properties of two cosmic models based on Yang-Mills gravity:

(I) The Expansion-Contraction (EC) model with energy conservation,

(II) A permanent expansion model without energy conservation.

5-10. Expansion-Contraction picture of the Big Jets model

The previous discussion of the EC model can be applied to both the evolution of a matter half-universe and that of an antimatter half-universe separately. The picture of expansion and contraction is particularly simple for observers in the center-of-mass frames of those two half-universes. However, what happens to the equation of motion for the entire universe? For a highly simplified Big Jets model, we postulate the effective metric tensor $G_{\mu\nu}(r)$ in the super-macroscopic limit in Yang-Mills gravity to be given by (5.4) with

$$J_{\mu\nu} = \eta_{\mu\nu}R(r), \qquad G_{\mu\nu}(r) = \eta_{\mu\nu}R^2(r). \tag{5.46}$$

We consider this to be the basic effective metric tensor for the motion of the matter half-universe for all epochs. Each of the two half-universes is idealized as a mass point located at its center-of mass. The trajectory of their motion is given by the geodesic of the metric, $ds^2 = G_{\mu\nu}(r)dx^\mu dx^\nu$ with the Sudarshan scale factor $R(r)$ in (5.46) and (5.13). Thus, the previous results (5.13)–(5.19) can be used for the evolution of either of the half-universes, provided the galactic mass M and its constant energy E in (5.18) are respectively replaced by the mass M_m and the energy E_m of a half-universe. CPT invariance implies that the mass and energy of the antimatter half-universe are the same as those of the matter half-universe. Thus, instead of R_o in (5.18), we have

$$R_{om} = \frac{E_m^2}{M_m^2 c^4 b_o^2} \approx \frac{E_m^2}{M_m^2 c^4} 10^{10} > 10^{10}\ lightyear, \tag{5.47}$$

Table 4.1   Expansion-Contraction and Expansion Models

(I) Expansion-Contraction Model	(II) Permanent Expansion Model
metric tensor: $G_{\mu\nu}(r) = \eta_{\mu\nu}R^2(r)$, Sudarshan scale factor $R(r) \propto r^{1/2}$	$G_{\mu\nu}(t) = \eta_{\mu\nu}U^2(t)$, where $U(t) \propto t^{1/2}$
energy is conserved in galactic motion	energy is not conserved
expansion: $T_o > t > 0$, $R_o > r > r_o$	$\infty > t > 0$, $\infty > r > 0$
contraction: $2T_o > t > T_o$, $R_o > r > r_o$	(no correspondence)
radius of EC cosmos: $R_o > 10^{10}$ ly	radius of expan. cosmos$\rightarrow \infty$
complete EC cycle: $2T_o > 2 \times 10^{10}$ y	$cosmos\ \ lifetime\ \rightarrow\ \infty$
expansion radius: $r(t) = R_o - \frac{c^2(T_o-t)^2}{4R_o}$	$r(t) = \frac{2c}{\Omega^2}\sqrt{1+\Omega^2 t}$
expansion velocity: $\dot{r} = c\sqrt{1 - r/R_o}$	$\dot{r} = c/\sqrt{1+\Omega^2 t} > 0$
contraction velocity: $\dot{r} = -c\sqrt{1 - r/R_o}$	(no correspondence)
expan. deceleration: $\ddot{r} = -c^2/(2R_o)$	$\approx -c/(2\Omega t^{3/2})$, $m >> p$
Okubo $eq : G^{\mu\nu}(r)\partial_\mu S \partial_\nu S - M^2 = 0$	$G^{\mu\nu}(t)\partial_\mu S \partial_\nu S - M^2 = 0$.
Okubo soln for initial velocity: $\dot{r} \rightarrow c$	initial velocity: $\dot{r} \rightarrow c$
Okubo soln for initial radius: $r(0) = r_o > 0$	$r(0) = 0$
Okubo soln for initial cosmic redshift: $z = \infty$	$z = \infty$
Okubo soln for final velocity: $\dot{r} \rightarrow 0$	$final\ velocity :\ \dot{r} \rightarrow 0$
time for contraction: $2T_o \geq t \geq T_o$	(no correspondence)
contraction radius: $r(t) = R_o - \frac{c^2}{4R_o}(t - T_o)^2$	(no correspondence)
contraction vel: $\dot{r}(t) = \frac{-c^2}{2R_o}(t - T_o) < 0$,	(no correspondence)
contraction acc: $|\ddot{r}| = |\frac{-c^2}{2R_o}| < |-4.5 \times 10^{-10}\frac{m}{s^2}|$	(no correspondence)

for the size or the radius R_{om} of the matter (or antimatter) half-universe in the Big Jets model, where we assume the half-universe to be roughly a sphere. So far, there are no data to calculate the unknown parameters E_m and M_m.

It is natural to assume that the super-macroscopic effective metric tensor $G_{\mu\nu}$ with the Sudarshan scale factor $R(r)$ in (5.3) and (5.13) holds for the Big Jets model. Thus the action and the Lagrangian of the Big Jets model are the same as those in (5.14). In this Big Jets model, let us assume a simple initial condition: Namely, the two centers-of-mass of the matter and antimatter half-universes are located near the origin of the inertial frame F_{tu} for the total universe. One can imagine that the centers of these two half-universes will move away from each other in opposite directions due to the CPT theorem and the conservation of momentum. The motion of the matter half-universe (which is idealized as a mass point) is determined by the Okubo equation of motion (5.45), or the Lagrange Eq. (5.15), in which the mass M of a galaxy in the Lagrangian is replaced by the total mass M_m of the matter half-universe.

Based on Lagrangian dynamics with Yang-Mills gravity for a super-macroscopic system, the Lagrangian L_{BJ} in the Big Jets model is

$$L_{BJ} = -M_m c^2 R(r)\sqrt{1 - \dot{r}^2/c^2}. \tag{5.48}$$

This Lagrangian leads to the conserved energy E_m in the motion of the matter half-universe

$$E_m = M_m R(r)c^2/\sqrt{1 - \dot{r}^2/c^2} = constant, \tag{5.49}$$

because the Big Jets Lagrangian (5.48) is time-independent, just like those in (5.14)–(5.16).

The discussions and derivations of equations for galaxies in Secs. 4-2 and 4-3 can be applied to the present case for the motion of the half-universes, modeling them as spheres, each with a maximum radius R_{om}. Equation (5.49) gives similar expansion and contraction velocities (5.19) with the maximum radius R_o of a galaxy replaced by the maximum radius R_{om} of the matter half-universe:

(I) The velocity $V_m = \dot{r}$ of the matter half-universe and its radial position $r(t)$ observed in the frame F_{tu} are as follows:

$$expansion: \quad V_m = \dot{r} = c\sqrt{1 - r/R_{om}}, \qquad 0 < t < T_{om}, \tag{5.50}$$

$$contraction: \quad V_m = \dot{r} = -c\sqrt{1 - r/R_{om}}, \qquad T_{om} < t < 2T_{om},$$

$$r(t) = R_{om} - \frac{c^2}{4R_{om}}(t - T_{om})^2, \quad T_{om} = \frac{2R_{om}}{c}\sqrt{1 - r(0)/R_{om}},$$

where R_{om} is given by (5.47). It is understood that r, $\dot{r}$, $r(0)$, etc. in (5.50) refer to quantities measured from the frame of the total universe F_{tu}, which can be pictured as the inertial frame whose origin is at rest at between the centers-of-mass of the matter and antimatter half-universes.

At time $t > 0$, the two Big Jets may be pictured as two gigantic fireballs moving in opposite directions. Observers in the inertial frame F_{tu} for the total universe will measure the velocity $\dot{r}'$ of the antimatter half-universe in the opposite direction of $\dot{r}$.

(II) The velocity $V_a = \dot{r}' = -\dot{r}$ of the antimatter half-universe and its radial position $r'(t)$ observed in the frame F_{tu} are

$$\text{expansion}: \ V_a = -c\sqrt{1 - r'/R_{om}}, \quad 0 < t < T_o, \ r' > 0, \quad (5.51)$$

$$\text{contraction}: \ V_a = +c\sqrt{1 - r'/R_{om}}, \quad T_o < t < 2T_o, \ r' > 0,$$

$$r'(t) = R_{om} - \frac{c^2}{4R_{om}}(t - T_{om})^2, \quad T_{om} = \frac{2R_{om}}{c}\sqrt{1 - r'(0)/R_{om}}.$$

Of course, the Big Jets model under consideration is a highly simplified model. For the motion of the two jets, we simply assume that each jet can be viewed as a gigantic fireball with all masses concentrated at its center-of-mass, which is not affected by the expansion of the matter and the antimatter half-universes. At the super-macroscopic scale, the motion of classical objects is naturally assumed to be dictated by the effective cosmic metric tensor (5.46) with the Sudarshan scale factor $R(r)$ given by (5.13) or the Okubo Eq. (5.45).

For a simple picture of the evolution of the total universe based on (5.14)–(5.19), we assume that all galaxies have roughly the same mass and that the constant effective energy E in (5.17) is also roughly the same for all galaxies (or anti-galaxies) in the matter (or antimatter) half-universe. Consequently, we have the simple relations,

$$\frac{M_m}{E_m} \approx \frac{M}{E}, \quad R_{om} \approx R_o, \quad T_{om} \approx T_o, \quad (5.52)$$

where we have used (5.20) for R_o related to the motion of galaxies and (5.47) for R_{om} for the motion of the matter and the antimatter half-universes. These results are consistent with the assumption that the motion of all these super-macroscopic objects obeys the geodesic of the same effective cosmic metric tensor $G_{\mu\nu}(r)$ in (5.3) and (5.13).

The results (5.50)–(5.52) and (5.34) based on Yang-Mills gravity, the cosmic isotropy principle and the Lagrangian dynamics (or the cosmic Okubo equation) suggest an interesting picture for the annihilation and creation of a Big Jets physical universe:

At the end of the EC cycle at time $t = 2T_o$, all galaxies move back to their original positions $r(0)$. At the same time, the two centers-of-mass of the matter half-universe and the antimatter half-universe also move with high energies to near the origin of the inertial frame F_{tu} of the total universe with speeds close to the speed of light ($\dot{r}(2T_o) \approx -c$ because $r(0) <<< R_o$ in (5.40)). Broader Particle-Cosmology implies that when all these particles and antiparticles of the whole universe rush back to the vicinity of the origin of the frame F_{tu}, unimaginably complicated interactions will take place. Very roughly speaking, the two matter and antimatter half-universes will annihilate each other and naturally cause the birth of a new baby universe.

How can a human being with a lifetime of roughly 100 years find evidence for a once in approximately 100 billion years cosmic rebirth? The evolution of the universe could be one of the biggest problems for curious physicists of the future to ponder based on Broader Particle-Cosmology. In this connection, we note that a detailed analysis and test of the prediction of the general Hubble's recession velocity in (5.26) and (5.27) by the EC model could help find an answer. For example, when there is good enough data on recession velocities and cosmic redshifts, one could have experimental tests of the square root $\sqrt{1 - r/R_o}$ in (5.19) or the modified Hubble's law (5.27). It would be difficult to test positive and negative signs for the radial velocity $\dot{r}$.

To conclude Part I, our biggest hope is that the Big Jets model based on the fundamental CPT invariance of particle physics could be tested by searching for a very small hemispheric anisotropy[17, 18] in the CMB radiation due to the existence of a distant antimatter half-universe.

References

1. J. P. Hsu, L. Hsu, D. Katz, Mod. Phys. Lett. A **33** 1850116 (2018). A. Friedmann, Z. Phys. **10**, 377 (1922); G. Lematre, Ann. Soc. Sci. de Bruxelles (in French), **A47**, 49 (1927); H. P. Robertson, Astrophys. J. **82**, 284 (1935); A. G. Walker, Proc. Lond. Math. Soc. **42**, 90 (1936).
2. J. P. Hsu and L. Hsu, *Space-Time, Yang-Mills Gravity and Dynamics of Cosmic Expansion* (World Scientific, 2020) pp. 113–127, pp. 157–165, p. 217, p. 241, p. 269, p. 284. For solving the Okubo Eq. (5.17), see p. 232.

3. E. P. Wigner, *Symmetries and Reflections, Scientific Essays* (The MIT Press, 1967) pp. 52–53; S. S. Chern (and W. H. Chen), *Lecture Notes on Differential Geometry* (Lian Jing Publisher, 1987) p. iv. Wigner wrote: 'Evidently, the usual statements about future positions of particles, as specified by their coordinates, are not meaningful statements in general relativity. This is a point which cannot be emphasized strongly enough ....' See also Ref. 3 in Ch. 1.

4. J. P. Hsu, Eur. Phys. J. Plus **126**: 24 (2011). DOI:10.1140/epjp/i2011-11024-x.

5. F. Dyson, in *100 Years of Gravity and Accelerated Frames: The Deepest Insights of Einstein and Yang-Mills.* (Eds. J. P. Hsu and D. Fine, World Scientific, 2005) p. 348.

6. Presented in the *14th International Conf. on Gravitation, Astrophysics and Cosmology*, L. Hsu and J. P. Hsu, 'Dynamic Models for the Beginning, Hubble Law and the Future of the Universe Based on Strong Cosmological Principle and Yang-Mills Gravity' (Aug. 2020).

7. L. Hsu and J. P. Hsu, Chin. Phys. C **43** No. 10, 105103 (2019).

8. S. Weinberg, *Cosmology* (Oxford, 2008) pp. 34–47, p. 510. (See p. 39 for flatness problem.)

9. M. Khan, Y. Hao and J. P. Hsu, EPJ Web of Conferences 168, 04004 (2018).

10. I. E. Segal, J. F. Nicoll, P. Wu and Z. Zhou, Astrophysical. J., 411, 465 (1993).

11. J. P. Hsu and L. Hsu, Nuovo Cimento, **112**, 575 (1997).

12. E. Hubble, Proc. Nat. Acad. Sci. (PNAS) **15**, 168 (1929).

13. T. D. Lee, *Particle Physics and Introduction to Field Theory* (Harwood Academic Publishers, 1981), pp. 746–752.

14. A. Gamow, Nature **162** (4122) 680 (1948).

15. H. Y. Chiu, Phys. Rev. Letter. 17, 712, (1966);

16. Ya. B. Zel'dovich, Advances Astron. Astrophys. 3 242 (1965).

17. See, for example, G. F. Smoot and P. M. Lubin, Astrophys. J. Part 2- Letters to the Editor, vol. 234, Dec. 1, 1979; DOI: 10.1086/183114. The experiment measured the large-angular-scale anisotropy in the cosmic background radiation from the Southern Hemisphere.

18. M. Planck, ESA Science and Technology, 'HEMISPHERIC ASYMMETRY AND COLD SPOT IN THE COSMIC MICROWAVE BACKGROUND'. Google search.

Part II

Symmetry-Unified Quark-Cosmic Model
Based On General Yang-Mills Symmetry

6

A Universal Principle of Interactions for Quarks and Leptons

Dynamic Symmetry Is The Heartbeat Of The Universe

6-1. The universal principle of general Yang-Mills symmetry

Symmetries in physics can be divided into two classes, dynamic symmetries and geometric symmetries. Dynamic symmetries such as U_1 or SU_{3c} dictate both interactions and conservation laws, while geometric symmetries such as the Poincaré group dictate only conservation laws, as noted by Wigner.[1]

We call the fundamental postulate for the interactions of quarks, leptons and gauge fields the 'Universal Principle of General Yang-Mills Symmetry' and it is:

All dynamic symmetries with conserved 'charges' are dictated by general Yang-Mills symmetry with (Lorentz) vector gauge functions.

General Yang-Mills (YM) symmetry groups include both internal groups such as U_1 and SU_3 and the external space-time translational group T_4 associated with Yang-Mills gravity. By definition, these include both Abelian and non-Abelian groups.

General YM symmetry is associated with a new characteristic phase function $P(x)$ rather than the usual scalar phase (or gauge) function $\Lambda(x)$ in gauge transformations. The characteristic phase function is an action integral with a vector function $\omega_\lambda(x)$ and is a Hamilton's characteristic function.[2] It involves a fixed initial point, a variable end point, and the path satisfies a Lagrange equation.

The electroweak gauge groups $SU_2 \times U_1$ with spontaneous symmetry breaking can be considered to be general Yang-Mills symmetry groups with special vector gauge functions, which can be expressed in terms of space-time derivatives of scalar gauge functions, i.e., $\omega_\mu^a(x) = \partial_\mu \omega^a(x)$, so that they have the usual gauge field equations.

The phase function in the usual gauge transformation of a fermion field can also be generalized to become a characteristic phase function, which is an action-integral,

$$phase\ function: \quad P(x) = \left(\int_{x'_o}^{x'_e=x} \omega_\mu(x')dx'^\mu \right)_{Le}.$$

As noted previously, it involves vector gauge functions, $\omega_\mu(x') = \omega^a(x')L^a$ with $x' \equiv x'^\mu$. For an internal group with generators L^a, it is defined to have a fixed initial point x'_o and a variable end point $x'_e = x$, and the path is required to satisfy a Lagrange equation (Le) to be specified later. This new phase function $P(x)$ is the same as a Hamilton's characteristic function[2,3] and it is mathematically well defined as the local function. Generalized gauge invariant Lagrangians with internal gauge groups involve new gauge fields that satisfy fourth-order differential equations and will be called 'phase fields.' However, the usual gauge fields still satisfy the usual second-order equations.

In a special subset of gauge functions in which the vector gauge functions can be expressed as the space-time derivative of a scalar function $\omega(x)$, i.e., $\omega_\mu(x) = \partial_\mu\omega(x)$, the new phase function $P(x)$ simplifies to the usual phase function and general Yang-Mills symmetry reduces to the usual gauge symmetry. In this sense, the generalized gauge symmetry includes the usual gauge symmetry as a special case. It is interesting to note that (Lorentz) vector gauge functions also appear in the gauge transformations of the external translational gauge symmetry in Yang-Mills gravity based on flat space-time.[4] Such an external gauge group with a vector gauge function is still associated with the second-order field equations.

In 1955, Lee and Yang suggested the existence of a U_1 gauge field associated with the conservation law of baryon number (or charge).[5] Their theory is U_1 gauge invariant and is formally identical to the electromagnetic theory. They used Eötvös' experiment to estimate the coupling strength between nucleons and found the new inverse-square force between baryon charges to be much weaker ($\leq 10^{-5}$) than the gravitational force. Thus, there are no observable physical effects whatsoever in high energy laboratory related to such an extremely weak baryon charge.

In 2005, a modified Lee-Yang gauge field with a fourth-order field equation was discussed within the framework of the usual baryonic U_{1b} gauge symmetry[6] using a new Lagrangian involving the quadratic form of the derivatives of the gauge curvature, $\partial^\alpha B_{\mu\nu}$. This was motivated by the desire to understand the late-time accelerated expansion of the universe on the basis of the baryonic gauge field associated with the established law of conservation of baryon number, rather than an ad hoc 'dark energy.'[a] However, the modified Lagrangian of the baryonic gauge field with U_{1b} symmetry is not unique in this approach.[6]

Therefore, it is natural to ask whether there is a new gauge symmetry that can 'uniquely' specify an invariant Lagrangian in analogy to the usual gauge symmetry, so that we can have a gauge symmetry foundation for the fourth-order gauge field equations. The answer turns out to be in the affirmative and this new gauge-symmetric framework (i.e., 'general Yang-Mills symmetry') appears to be logically as sound as the usual gauge fields with the second-order gauge field equations.

In the literature, there have been many discussions related to higher-order field equations and their quantizations. For example, field equations derived from Lagrangians with higher derivatives were investigated by J. S. de Wet in 1948.[7] He showed how such field equations can be put into Hamiltonian form and how the quantization of boson and fermion fields can be carried out. He established that quantization is relativistically invariant and consistent with the field equations. However, he also found that the Hamiltonian proves to be different, in general, from the integral of the 00 component of the energy momentum tensor. Pais and Uhlenbeck also discussed Lagrangians involving higher order derivatives.[8] They found that there is no essential problem, except that dynamical systems have non-definite energy.

Various aspects of higher-order field equations have also been discussed in the literature.[9–11] However, it seems that no connection has been made to basic physical fields[b] and no experimental predictions have been made based on higher-order field equations. Some have expressed the hope that higher-order field equations might help eliminate ultraviolet divergences in quantum field theory. However, this hope turned out to be difficult to realize because once the ultraviolet divergences are eliminated in this way, the unitarity of the S-matrix will, in general, be upset so that the probability for a certain physical process may be negative, making the theory unphysical.

The new general Yang-Mills symmetry appears to be a deeper-lying symmetry principle because it includes the usual gauge symmetry, Abelian

---

[a]As noted by J. M. Keynes: 'The difficulty lies not so much in developing new ideas as in escaping from old ones.'
[b]In mechanical vibrations, the beam displacement can be described by fourth-order differential equations. We thank E. C. G. Sudarshan for the comment.

or non-Abelian, as a special subset of gauge functions (i.e., $\omega_\mu^a(x) = \partial_\mu \omega^a(x)$). In particular, the general U_1 symmetry predicts a new kind of cosmic linear potential produced by baryonic gauge fields, which could provide an understanding of the late-time accelerated cosmic expansion. Furthermore, the non-definite energies of quanta associated with the fourth-order field equations are not observable because these quanta are permanently confined in the physical systems. They do not propagate as free particles and hence, cannot be observed. The fourth-order equations are only for gauge fields and by contrast, matter fields (or fermion fields) still satisfy the first-order equations. Therefore such field theories with general Yang-Mills symmetries do not contradict experiments.

Within the framework of general Yang-Mills symmetry, we use the term 'phase fields' to denote these new gauge fields with fourth-order differential equations to distinguish them from the usual gauge fields with second-order equations and from other formulations of field theories with higher-order equations.

6-2. General U_1 Yang-Mills symmetry

The general U_1 transformations for quarks $q(x)$ are given by[12]

$$q'(x) = (1 - iP)q(x), \quad \bar{q}'(x) = \bar{q}(x)(1 + iP), \qquad (6.1)$$

where P is Hamilton's characteristic phase function,

$$P = \left(g_b \int_{x_o'}^{x} dx'^\mu \Lambda_\mu(x') \right)_{Le}, \qquad (6.2)$$

and Λ_μ are arbitrarily infinitesimal vector functions. The general Yang-Mills (YM) transformations for the general U_1 phase fields $B_\mu(x)$ are given by

$$B'_\mu(x) = B_\mu(x) + \Lambda_\mu(x). \qquad (6.3)$$

To see the equation Le in (6.2), let us consider the variation of P with a variable end point and a fixed initial point x_o'. We have

$$\delta P = g_b \frac{\partial L}{\partial \dot{x}^\lambda} \delta x^\mu + g_b \left(\int_{\tau_o}^{\tau} \left(-\frac{d}{d\tau} \frac{\partial L}{\partial \dot{x}^\lambda} + \frac{\partial L}{\partial x^\lambda} \right) \delta x^\lambda d\tau \right)_{Le} \qquad (6.4)$$

where we write (6.2) in the usual form of a Lagrangian L with the help of a parameter τ,

$$P = \left(g_b \int_{\tau_o}^{\tau} L \, d\tau \right)_{Le}, \quad L = \dot{x}^\mu \Lambda_\mu(x), \quad \dot{x}^\mu = \frac{dx^\mu}{d\tau}. \qquad (6.5)$$

We require that the paths in (6.4) are those satisfy the Lagrange equation Le, i.e.,

$$-\frac{d}{d\tau}\frac{\partial L}{\partial \dot{x}^\lambda} + \frac{\partial L}{\partial x^\lambda} = 0, \qquad L = \dot{x}^\mu \Lambda_\mu(x). \tag{6.6}$$

This Lagrange equation can also be written in the form

$$\frac{d\Lambda_\lambda}{d\tau} - \frac{\partial \Lambda_\mu}{\partial x^\lambda}\dot{x}^\mu = 0. \tag{6.7}$$

Thus, the integral in (6.4) vanishes due to (6.6) and we have the relation[3]

$$\partial_\mu P = g_b \Lambda_\mu(x). \tag{6.8}$$

As usual, the general U_1 covariant derivatives are defined as

$$\Delta_\mu = \partial_\mu + ig_b B_\mu. \tag{6.9}$$

The general U_1 curvatures $B_{\mu\nu}$ are given by

$$[\Delta_\mu, \Delta_\nu] = ig_b B_{\mu\nu}, \tag{6.10}$$

$$B_{\mu\nu} = \partial_\mu B_\nu - \partial_\nu B_\mu. \tag{6.11}$$

The general U_1 curvature $B_{\mu\nu}$ is not gauge invariant under the gauge transformation (6.3) because

$$B'_{\mu\nu} = B_{\mu\nu} + \partial_\mu \Lambda_\nu(x) - \partial_\nu \Lambda_\mu(x). \tag{6.12}$$

However, we have the following general Yang-Mills (gYM) transformations for $\partial^\mu B_{\mu\nu}(x)$, and $\bar{q}\Delta_{b\mu}q$:

$$\partial^\mu B'_{\mu\nu}(x) = \partial^\mu B_{\mu\nu}(x) + \partial^\mu\{\partial_\mu \Lambda_\nu(x) - \partial_\nu \Lambda_\mu(x)\} = \partial^\mu B_{\mu\nu}(x), \tag{6.13}$$

$$\bar{q}'\gamma^\mu \Delta'_\mu q' = \bar{q}\gamma^\mu \Delta_\mu q, \tag{6.14}$$

provided the restrictions

$$\partial^\mu\{\partial_\mu \Lambda_\nu(x) - \partial_\nu \Lambda_\mu(x)\} = 0 \tag{6.15}$$

are imposed for (6.13) to hold.

In this way, we can use the invariant 'phase curvature' $\partial^\mu B_{\mu\nu}$ to formulate a field theory with general Yang-Mills U_1 symmetry.

The general U_1 group properties can be demonstrated. For group properties of the general U_1 transformations (6.1)–(6.3), we have the inverse, identity and associativity for group operations. To see the closure property, we consider two consecutive transformations involving infinitesimal $\Lambda_{r\mu}$ and $\Lambda_{s\mu}$. We have

$$(1 - iP_r)(1 - iP_s) \approx (1 - iP_t), \qquad P_t = P_r + P_s$$

$$P_t = \left(g_b \int_{x_o'}^{x_e'=x} dx'^\mu \left[\Lambda_{t\mu}(x') \right] \right)_{Le(t)} = P_t(\Lambda_t, x) \qquad (6.16)$$

$$\Lambda_{t\mu}(x') = \Lambda_{r\mu}(x') + \Lambda_{s\mu}(x').$$

The Lagrange equation $Le(t)$ in (6.16) is given by (6.6) with $L = \dot{x}^\mu \Lambda_{t\mu}(x)$.

After some straightforward calculations, the results in (6.4)–(6.8) are still true for two consecutive transformations, where P and Λ_μ are respectively replaced by P_t and $\Lambda_{r\mu} + \Lambda_{s\mu}$. Furthermore, the result (6.13) with $\Lambda(x)$ replaced by $\Lambda_t(x)$ holds, provided we impose the constraint (6.15) with $\Lambda_\mu(x)$ replaced by $\Lambda_{r\mu}(x) + \Lambda_{s\mu}(x)$. Thus, the general transformations (6.13)–(6.14) together with the constraint (6.15) can be considered general U_1 transformations with the required group properties.

6-3.　General SU_N symmetry

The discussions of the general U_{1b} group can be applied to the non-Abelian group SU_N, which includes color SU_{3c} as a special case. The general SU_N transformations for quarks $q(x)$ are given by[13]

$$q'(x) = (1 - iP)q(x), \quad \bar{q}'(x) = \bar{q}(x)(1 + iP), \qquad (6.17)$$

where P is an infinitesimal phase functional (or action integral),

$$P = L^a \left(g_s \int_{x_o'}^{x} dx'^\mu \omega_\mu^a(x') \right)_{Le} = L^a P^a(\omega, x), \qquad (6.18)$$

$$[L^a, L^b] = if^{abc} L^c, \qquad (6.19)$$

where $\omega = \omega_\mu^a$ are arbitrary vector functions and L^a are SU_N generators.[14] The general Yang-Mills (gYM) transformations for the SU_N gauge fields, $H_\mu^a(x)$ are given by

$$H_\mu'(x) = H_\mu(x) + \omega_\mu(x) - i[P(x), H_\mu(x)], \quad H_\mu = H_\mu^a L^a, \qquad (6.20)$$

or

$$H_\mu'^a(x) = H_\mu^a(x) + \omega_\mu^a(x) + g_s f^{abc} P^b(\omega, x) H_\mu^c(x), \qquad (6.21)$$

where $P^b(\omega, x)$ is defined in (6.18). To see the equation Le in (6.18), let us consider the variation of P with a variable end point and a fixed initial point x_o'. We have

$$\delta P = g_s \frac{\partial L}{\partial \dot{x}^\lambda} \delta x^\mu + g_s \left(\int_{\tau_o}^{\tau} \left(-\frac{d}{d\tau} \frac{\partial L}{\partial \dot{x}^\lambda} + \frac{\partial L}{\partial x^\lambda} \right) \delta x^\lambda d\tau \right)_{Le} \qquad (6.22)$$

where we write (6.18) in the usual form of a Lagrangian with the help of a parameter τ,

$$P = \left(g_s \int_{\tau_o}^{\tau} L \, d\tau\right)_{Le},\qquad(6.23)$$

$$L = \dot{x}^\mu \omega_\mu^a(x) L^a, \quad \dot{x}^\mu = \frac{dx^\mu}{d\tau}.$$

We require that the paths in (6.22) are those that satisfy the Lagrange equation Le, i.e.,

$$-\frac{d}{d\tau}\frac{\partial L}{\partial \dot{x}^\lambda} + \frac{\partial L}{\partial x^\lambda} = 0, \qquad L = \dot{x}^\mu \omega_\mu^a(x) L^a.\qquad(6.24)$$

This Lagrange equation can also be written in the form

$$\frac{d\omega_\lambda^a}{d\tau} - \frac{\partial \omega_\mu^a}{\partial x^\lambda}\dot{x}^\mu = 0,\qquad(6.25)$$

similar to Eq. (6.7) for the general U_1 symmetry. Thus, the integral in (6.22) vanishes and we have the relation[8]

$$\partial_\mu P = g_s \omega_\mu^a(x) L^a = g_s \omega_\mu.\qquad(6.26)$$

In the following discussions, we shall concentrate on the specific case $N = 3$. As usual, the SU_{3c} gauge covariant derivatives are defined as

$$\Delta_\mu = \partial_\mu + ig_s H_\mu^a \frac{\lambda^a}{2}, \quad L^a = \frac{\lambda^a}{2},\qquad(6.27)$$

where λ^a are the 3×3 Gell-Mann matrices.[14] The SU_{3c} gauge curvatures $H_{\mu\nu}^a$ are given by

$$[\Delta_\mu, \Delta_\nu] = ig_s H_{\mu\nu},\qquad(6.28)$$

$$H_{\mu\nu} = \partial_\mu H_\nu - \partial_\nu H_\mu + ig_s[H_\mu, H_\nu],\qquad(6.29)$$

or

$$H_{\mu\nu}^a = \partial_\mu H_\nu^a - \partial_\nu H_\mu^a - g_s f^{abc} H_\mu^b H_\nu^c.$$

It follows from Eqs. (6.20)–(6.29) that we have the following gYM transformations for $\partial^\mu H_{\mu\nu}(x)$, and $\bar{q}\Delta_{b\mu}q$:

$$\partial^\mu H_{\mu\nu}'(x) = \partial^\mu H_{\mu\nu}(x) - [P(x), \partial^\mu H_{\mu\nu}(x)]\qquad(6.30)$$

$$\bar{q}'\gamma^\mu \Delta_\mu' q' = \bar{q}\gamma^\mu \Delta_\mu q,\qquad(6.31)$$

provided the restrictions

$$\partial^\mu \{\partial_\mu \omega_\nu(x) - \partial_\nu \omega_\mu(x)\} + ig_s [\omega^\mu(x), H_{\mu\nu}(x)] = 0 \qquad (6.32)$$

are imposed for (6.30) to hold. This constraint is similar to that for gauge functions of Lie groups in the conventional non-Abelian gauge theories.[15]

6-4.   Consecutive general Yang-Mills transformations

Now let us consider the general SU_N group properties. For group properties of the general SU_N transformations (6.17) and (6.20), it is convenient to consider the infinitesimal transformation, $e^{-iP} \approx (1 - iP)$. Clearly, we have inverse, identity and associativity for group operations. To see the closure property, we consider two consecutive transformations involving $\omega_{r\mu}^a$ and $\omega_{s\mu}^a$.[13] We have

$$(1 - iP_r)(1 - iP_s) = (1 - iP_t), \qquad P_t = P_r + P_s$$

$$P_t = \left(g_s \int_{x'_o}^{x'_e = x} dx'^\mu \left[\omega_{t\mu}^a(x')L^a \right] \right)_{Le(t)} = P_t(\omega_t, x) \qquad (6.33)$$

$$\omega_{t\mu}^a(x')L^a = \omega_{r\mu}^a(x')L^a + \omega_{s\mu}^a(x')L^a.$$

The Lagrange equation Le in (6.18) is given by (6.24) with

$$L = \dot{x}^\mu \omega_{t\mu}^a(x)L^a.$$

After some tedious but straightforward calculations, the results in (6.22)–(6.26) are still true for two consecutive transformations, where P and $\omega_\mu^a L^a$ are respectively replaced by P_t and $\omega_{r\mu}^a L^a + \omega_{s\mu}^b L^b$. Furthermore, the result (6.33) with $P(x)$ replaced by P_t holds, provided we impose the constraint (6.32) with $\omega_\mu(x)$ replaced by $\omega_{r\mu}(x) + \omega_{s\mu}(x)$. Thus, the general transformations (6.33)–(6.36) together with the constraint (6.32) can be considered general SU_N transformations with the required group properties.

For group properties related to general Yang-Mills symmetry, let us consider two consecutive general Yang-Mills transformations for the SU_N gauge fields,

$$H_\mu(x) = H_{\mu\nu}^a L^a.$$

We have the transformations,

$$H_\mu''(x) = H_\mu'(x) + \omega_{r\mu}(x) - i[P_r, H_\mu'(x)], \qquad (6.34)$$

$$H'_\mu(x) = H_\mu(x) + \omega_{s\mu}(x) - i[P_s, H_\mu(x)], \qquad (6.35)$$

$$H''_\mu = H''^a_\mu L^a, \qquad \omega_{r\mu} = \omega^a_{r\mu} L^a.$$

Based on these transformations, we obtain the following relations

$$H''_\mu = H_\mu + \omega_{t\mu} - i[P_t, H_\mu], \qquad (6.36)$$

$$\omega_{t\mu} = \omega_{r\mu} + \omega_{s\mu}, \qquad P_t = P_r + P_s,$$

where P_t can be expressed as

$$P_t = \left(g \int_{x'_o}^{x} dx'^\mu \left[\omega^a_{r\mu}(x') L^a + \omega^b_{s\mu}(x') L^b \right] \right)_{Le(t)}, \qquad (6.37)$$

where $g \equiv g_s$. To see the equation $Le(t)$ in (6.37), let us consider the variation of P_t with a variable end point $x'_e = x$ and a fixed initial point x'_o. We have

$$\delta P_t = g \frac{\partial L_t}{\partial \dot{x}^\lambda} \delta x^\mu \qquad (6.38)$$

$$+ g \left(\int_{\tau_o}^{\tau} \left(-\frac{d}{d\tau} \frac{\partial L_t}{\partial \dot{x}^\lambda} + \frac{\partial L_t}{\partial x^\lambda} \right) \delta x^\lambda d\tau \right)_{Le(t)}$$

where we write (6.37) in the usual form of a Lagrangian with the help of a parameter τ,

$$P_t = \left(g \int_{\tau_o}^{\tau} L_t \, d\tau \right)_{Le(t)}, \qquad (6.39)$$

$$L_t = \dot{x}^\mu [\omega^a_{r\mu}(x) L^a + \omega^b_{s\mu}(x') L^b].$$

We require that the paths in (6.38) are those satisfy the Lagrange equation $Le(t)$, i.e.,

$$-\frac{d}{d\tau} \frac{\partial L_t}{\partial \dot{x}^\lambda} + \frac{\partial L_t}{\partial x^\lambda} = 0. \qquad (6.40)$$

Thus, the integral in (6.38) vanishes and we have the relation

$$\partial_\mu P_t = \omega^a_{r\mu}(x) L^a + \omega^b_{s\mu} L^b. \qquad (6.41)$$

The SU_N gauge covariant derivatives are defined as

$$\Delta_\mu = \partial_\mu + ig H^a_\mu L^a. \qquad (6.42)$$

The SU_N gauge curvatures $H_{\mu\nu} = H_{\mu\nu}^a L^a$ are given by

$$H_{\mu\nu} = \partial_\mu H_\nu - \partial_\nu H_\mu + ig[H_\mu, H_\nu]. \tag{6.43}$$

Under the general Yang-Mills transformations (6.18) and (6.20), we have

$$H_{\mu\nu}'' = \partial_\mu(H_\nu' + \omega_{r\nu} - i[P_r, H_\nu']) - \partial_\nu(H_\mu' + \omega_{r\mu} - i[P_r, H_\mu'])$$

$$+ig[(H_\mu' + \omega_{r\mu} - ig[P_r, H_\mu']), (H_\nu' + \omega_{r\nu} - ig[P_r, H_n'u])]$$

$$= \partial_\mu(H_\nu + \omega_{t\nu} - i[P_t, H_\nu]) - \partial_\nu(H_\mu + \omega_{t\mu} - i[P_t, H_\mu])$$

$$+ig[(H_\mu + \omega_{t\mu} - ig[P_t, H_\mu]), (H_\nu + \omega_{t\nu} - ig[P_t, H_\nu])]. \tag{6.44}$$

After cancellations of the terms $[H_\mu, \omega_{t\nu}]$ and $[H_\nu, \omega_{t\mu}]$, we obtain the following relations,

$$H_{\mu\nu}'' + \partial_\mu\omega_{t\nu} - \partial_\nu\omega_{t\mu} - ig[P_t, \partial_\mu H_\nu - \partial_\nu H_\mu] \tag{6.45}$$

$$-(ig)^2 \left([H_\mu, [P_t, H_\nu]] + [H_\nu, [H_\nu, P_t]]\right)$$

$$= H_{\mu\nu} - ig[P_t, H_{\mu\nu}] + \partial_\mu\omega_{t\nu} - \partial_\nu\omega_{t\mu}.$$

In contrast to the usual gauge symmetry, the SU_N gauge curvatures $H_{\mu\nu}$ do not transform according to the adjoint representation under two consecutive gYM transformations. However, $\partial^\mu H_{\mu\nu}$ has a proper transformation property,

$$\partial^\mu H_{\mu\nu}'' = \partial^\mu H_{\mu\nu} - ig[P_t, \partial^\mu H_{\mu\nu}], \tag{6.46}$$

provided the arbitrary infinitesimal vector gauge functions $\omega_{t\mu}$ satisfy the constraints

$$\partial^\mu(\partial_\mu\omega_{t\nu} - \partial_\nu\omega_{t\mu}) - ig[\omega_t^\mu, H_{\mu\nu}] = 0, \tag{6.47}$$

where $\omega_{t\mu} = \omega_{r\mu} + \omega_{s\mu}$. This result for two consecutive SU_N gauge transformations is consistent with the constraint in (6.32) for gYM transformations within the framework of general Yang-Mills symmetry.

References

1. E. Wigner, Proc. Nat. Acad. Sci. **51**, 5 (1964); H. Poincaré, Rendiconti del Circolo Matematics di Palermo **21**, 129 (1906), dated: Paris, July 1905. English translation, see H. M. Schwartz, Am. J. Phys. **39**, 1287 (1971); **40**, 862 (1972); E. Noether, Goett. Nachr., 235 (1918). English translation of Noether's paper by M. A. Tavel is online. Google search: M. A. Tavel, Noether's paper.
2. W. Yourgrau and S. Mandelstam, *Variational Principles in Dynamics and Quantum Theory*, (3rd ed, 1979, Dover), p. 50.
3. L. Landau and E. Lifshitz, *The Classical Theory of Fields* (Trans. by M. Hamermesh, Addison-Wesley, Cambridge, MA. 1951), p. 29.
4. R. Utiyama, in *100 Years of Gravity and Accelerated Frames, The Deepest Insights of Einstein and Yang-Mills* (Eds. J. P. Hsu and D. Fine, World Scientific, 2005), p. 157. See also T. W. B. Kibble, *ibid*, p. 168; C. N. Yang, *ibid*, pp. 527–538.
5. T. D. Lee and C. N. Yang, Phys. Rev. **98** 1501 (1955).
6. J. P. Hsu, Modern Phys. Lett. A **29** (06), 1450031(2014).
7. J. S. de Wet, Math. Proc. of the Cambridge Phil. Soc. **44**, 546 (1948).
8. A. Pais and G. E. Uhlenbeck, Phys. Rev. **79**, 145 (1950).
9. T. S. Chang, Proc. Roy. Soc. A **183**, 316 (1945).
10. F. Bopp, Ann. d. Physik, **38**, 345 (1940).
11. Y. Takano, Prog. Theor. Phys. **26**, 304 (1961); A. O. Barut and G. H. Muller, Annals of Phys. **20**, 203 (1962).
12. J. P. Hsu, Mod. Phys. Lett. A **31**, 1650200 (2016). Vector gauge functions and non-integral phase functions were used to investigate the form of gauge fields and wrapping number in gauge field theories. See J. P. Hsu, Phys. Rev. Lett. **36**, 1515 (1976).
13. J. P. Hsu, Chin. Phys. C **41**, 015101, Appendix, (2017).
14. K. Huang, *Quarks, Leptons and Gauge Fields*. (World Scientific, 1982). pp. 59–64, pp. 241–249.
15. B. W. Lee and Jean Zinn-Justin, Phys. Rev. **7**, 1047 (1973).

7

Finite Fermion Self-Masses and a Non-Propagating Phase Field

7-1.   Finite fermion self-masses

According to the universal principle of general Yang-Mills symmetry, the conserved baryonic charge of quarks could be associated with a general U_1 Yang-Mills symmetry. To show the new physical results of such a general U_1 symmetry, we demonstrate that the self-masses of these fermions under general U_1 dynamics are finite, in sharp contrast to the electron self-mass in the usual U_1 dynamics for electric charges.

To calculate the fermion self-masses, we concentrate on the quantum dynamics of quarks, whose baryonic charge g_b is associated with the general Yang-Mills U_1 symmetry. The Feynman-Dyson rules[a] for Feynman diagrams can be derived from the total Lagrangian L_{tot} including a quark Lagrangian L_{qU_1} and the gauge-fixing terms L_{gf},[1]

$$L_{tot} = L_{qU_1} + L_{gf}, \qquad (7.1)$$

$$L_{qU_1} = \frac{L_s^2}{2}\left(\partial^\mu B_{\mu\lambda}\partial_\nu B^{\nu\lambda}\right) + i\bar{q}(x)\gamma^\mu(\partial_\mu + ig_b B_\mu)q(x) - m_q\bar{q}(x)q(x),$$

$$L_{gf} = \frac{L_b^2}{2\xi}(\partial^\lambda\partial_\mu B^\mu)(\partial_\lambda\partial_\nu B^\nu), \qquad B_{\mu\nu} = \partial_\mu B_\nu - \partial_\nu B_\mu.$$

The gauge-fixing term L_{gf} is necessary only for the quantization of fields B_μ in (7.1).

We assume that fields with general U_1 symmetry in the Lorentz-Poincaré invariant Lagrangian (7.1) can be quantized, although the new fields $B_\mu(x)$ have non-trivial properties. We call the quanta of the phase fields $B_\mu(x)$ (which satisfy fourth-order field equations) 'phasons' for convenience and to stress their difference from the fields for the usual gauge bosons, which satisfy the second-order field equations. The phason propagator is related to the fourth-order field equations, $\partial^2\partial^2 B_\mu - (1 - 1/\xi)\partial^2\partial_\mu\partial_\nu B^\nu =$

---

[a]Dyson's rigorous derivation of the rules from a Lagrangian completed Feynman's original intuitive rules for all orders, made them applicable to all quantum field theories, and facilitated higher-order calculations.

$(g_b/L_s^2)\bar{q}\gamma_\mu q$, which can be derived from the Lagrangian (7.1). We find the phason propagator to be[2]

$$b_{\mu\nu}(k) = \frac{-i}{L_b^2(k^2 + i\epsilon)^2}\left[\eta_{\mu\nu} - (1 - \xi)\frac{k_\mu k_\nu}{k^2 + i\epsilon}\right].\tag{7.2}$$

The rule for the phason-quark 3-vertex $[\bar{q}(k_1)q(k_2)B_\mu(k_3)]$ is given by

$$-ig_b\gamma_\mu.\tag{7.3}$$

Other rules such as fermion propagators, a factor -1 for each fermion loop, etc. are the same as those in conventional QED.[1] An important difference from QED is that the phason propagator (7.2) has a better high-energy (or large k) behavior than the photon propagator. This property is a special feature of general Yang-Mills symmetry of the Lagrangian (7.1).

Let us consider new physical effects of the fourth-order field equations dictated by general U_1 symmetry in quantum dynamics. To see the divergence-free loop diagram, we consider the fermion self-mass δm. Following the Feynman-Dyson rules for writing the invariant amplitude $-iM_{fi}$, we have[1]

$$-i\delta m = (-ig_b)^2\int\frac{d^4k}{(2\pi)^4}b^{\mu\nu}(k)\frac{i\gamma_\mu[\gamma\cdot(p-k)+m]\gamma_\nu}{(p-k)^2 - m^2 + i\epsilon},$$

$$= \frac{-g_b^2}{L_s^2}[T_1 - (1-\xi)T_2].\tag{7.4}$$

The phason propagator $b^{\mu\nu}(k)$ is given by (7.2), where we have replaced k^2 by $k^2 + \lambda^2$ to avoid a possible infrared divergence. We shall take the limit $\lambda \to 0$ at the end of the calculations. Using dimensional regularization, we have

$$T_1 = \int_0^1 2x\,dx\int\frac{d^Dk}{(2\pi)^D}\frac{(2-D)\gamma\cdot(p-k)+Dm}{[k^2 - 2k\cdot p(1-x)+(p^2-m^2)(1-x)+\lambda^2x]^3},\tag{7.5}$$

where $D \to 4$ at the end of the calculations. After the shift $k_\mu \to k_\mu + p_\mu(1-x)$ and carrying out the dimensional regularization, we obtain

$$T_1 = \frac{-im}{8\pi^2}(2\alpha - \beta) = \frac{3g_b^2}{16\pi^2L_s^2m},\tag{7.6}$$

$$\alpha = \int_0^1\frac{x\,dx}{R} = \left[\frac{1}{2A}ln|R| - \frac{B}{2A}Z\right]_0^1,\quad R = Ax^2 + Bx + C,$$

$$\beta = \int_0^1\frac{x^2\,dx}{R} = \left[\frac{x}{A} - \frac{B}{2A^2}ln|R| + \frac{YZ}{2A^2}Z\right]_0^1.$$

As usual, it is understood that δm is sandwiched between free fermion spinors, so that one has $\gamma \cdot p = m$ and $p^2 = m^2$. Although the fermion self-mass is ultraviolet convergent, it may be infrared divergent. To keep it mathematically well-defined, we replace k^2 in (7.2) by $k^2 + \lambda^2$ with $\lambda^2 > 0$ and take the limit $\lambda \to 0$ at the end of the calculations.

Using the relation $1/(a^2 b) = \int_0^1 2x\,dx/[b + x(a - b)]^3$ and the Feynman gauge (gauge parameter $\xi = 1$ in (7.2) and (7.4)) for simplicity, we obtain the result

$$\delta m = \frac{-ig_b^2}{L_s^2} \int \frac{d^4k}{(2\pi)^4} \int_0^1 2x\,dx \frac{4m - 2\gamma \cdot (p - k)}{[k^2 - 2k \cdot p(1 - x) + \lambda^2 x]^3}$$

$$= \frac{-ig_b l^2}{L_s^2} \int_0^1 2x\,dx \frac{-i}{32\pi^2} \left[\frac{4m - 2x\gamma \cdot p}{m^2(1 - x)^2 - \lambda^2 x} \right]$$

$$= \frac{-4mg_b^2}{32\pi^2 L_s^2} (2A_1 - A_2), \tag{7.7}$$

where

$$A_1 = \int_0^1 \frac{x\,dx}{Ax^2 + Bx + C} = \left[\frac{1}{2A} ln|Ax^2 + Bx + C| - \frac{B}{2A} Z \right]_0^1, \tag{7.8}$$

$$A_2 = \int_0^1 \frac{x^2\,dx}{Ax^2 + Bx + C}$$

$$= \left[\frac{x}{A} - \frac{B}{2A^2} ln|Ax^2 + Bx + C| - \frac{B^2 - 2AC}{2A^2} Z \right]_0^1, \tag{7.9}$$

$$Z = \frac{1}{\sqrt{B^2 - 4AC}} ln \left[\frac{|2Ax + B - \sqrt{B^2 - 4AC}|}{|2Ax + B + \sqrt{B^2 - 4AC}|} \right]. \tag{7.10}$$

Since $A = C = m^2$, $B = -(2m^2 + \lambda^2)$, we have

$$A_1 = \frac{1}{2m^2} ln \frac{\lambda^2}{m^2} - \frac{1}{2m^2}, \tag{7.11}$$

$$A_2 = \frac{1}{m^2} + \frac{1}{m^2} ln \frac{\lambda^2}{m^2} - \frac{1}{2m^2}. \tag{7.12}$$

Thus, we obtain a finite fermion self-mass based on general Yang-Mills U_1 symmetry,

$$\delta m = \frac{3g_b^2}{16\pi^2 L_b^2 m}. \tag{7.13}$$

In comparison, the electron self-mass δm_e in the Feynman gauge based on conventional quantum electrodynamics (QED) has an ultraviolet divergence,[1,3]

$$\delta m_e = \frac{3e^2 m_e}{16\pi^2} ln\left(\frac{\Lambda^2}{m_e^2}\right), \tag{7.14}$$

where Λ is the ultraviolet cut-off.

It is interesting that the two infrared logarithmic divergent terms cancel each other in (7.7) in the Feynman gauge.

We also note that it is important to use the small parameter $\lambda^2 > 0$ to avoid divergences. If one were to use $\lambda^2 < 0$ in the calculation of the integrals A_1 and A_2 in (7.11) and (7.12), one would obtain an arctangent function rather than the logarithmic function Z in (7.10) and there would be two infrared divergent terms $\propto 1/\lambda$ in A_1 and A_2 that do not cancel each other.

7-2. The instantaneous 'phason' field and its Green's Functions

We assume that the phase fields can be quantized using path integrals to calculate the vacuum-to-vacuum amplitudes,[4] similar to the electromagnetic fields. The very intriguing properties of the phase field (or its quantum 'phason') can be seen more clearly by comparing them with the electromagnetic field (or its quantum 'photon'). In electrodynamics, in order to see the physical propagation properties of photons, one considers the time-independent Green's function in vacuum associated with the inhomogeneous Helmholtz wave equation,[5] i.e., $(\nabla^2 + k^2)G_k(R) = -4\pi\delta^3(\mathbf{R})$. One has, say, a divergent spherical wave from the source at the origin, $G_k^{(+)}(R) = e^{ikR}/R$. For the time-dependent Green's function, one has the well-known results,[5]

$$G^{(\pm)}(R, \tau) = \frac{1}{R}\delta(\tau \mp R), \tag{7.15}$$

$$R = |\mathbf{R}| = |\mathbf{r} - \mathbf{r}'|, \quad \tau = t - t'.$$

The argument in (7.15) implies that an effect observed at the position $\mathbf{r}$ at time t is caused by the action of a source at the position $\mathbf{r}'$ at an earlier time $t' = t - R$, where $c = 1$.

For the propagation property of the phase field $B_\mu(x)$ or the phason, we consider the invariant Green's function related to the phase field equation. Its invariant Green's function in vacuum takes the form

$$\partial_z^2 \partial_z^2 D(z) = \delta^4(z), \quad z^\mu = x^\mu - x'^\mu, \quad \partial_z^2 = \frac{\partial}{\partial z^\mu} \frac{\partial}{\partial z_\mu}, \tag{7.16}$$

for, say, the quark source. The formal solution of $\partial_x^2 \partial_x^2 B_\nu(x) = J_\nu(x)$ with the quark source $J_\nu(x)$ can be expressed as

$$B_\nu(x) = \int D(x - x') J_\nu(x') d^4 x'. \tag{7.17}$$

We define the Fourier transform $\overline{D}(k) \equiv \overline{D}(k_\sigma)$ of the Green's function $D(z) \equiv D(z^\lambda)$ by

$$D(z) = \frac{1}{(2\pi)^4} \int \overline{D}(k) e^{-ik_\mu z^\mu} d^4 k. \tag{7.18}$$

Since $\delta^4(z) = (2\pi)^{-4} \int d^4 k \, exp(-ik_\mu z^\mu)$, (7.16) and (7.18) lead to

$$\partial_z^2 \partial_z^2 D(z) = \frac{1}{(2\pi)^4} \int \overline{D}(k) k_\lambda^2 k_\sigma^2 e^{-ik_\mu z^\mu} d^4 k \tag{7.19}$$

$$= \frac{1}{(2\pi)^4} \int d^4 k \, e^{-ik_\mu z^\mu}.$$

Thus, we have

$$\overline{D}(k) = \frac{1}{k_\lambda^2 k_\sigma^2}, \quad k_\lambda^2 = k_\lambda k^\lambda = k_0^2 - \kappa^2, \quad \kappa = |\mathbf{k}|. \tag{7.20}$$

Using (7.18) and (7.19), we solve the Green's function $D(z)$ by carrying out the contour integration on the complex k_0 plane,[5]

$$D(z) = \frac{1}{(2\pi)^4} \int e^{-i\mathbf{k}\cdot\mathbf{z}} d^3 k \int_{-\infty}^{\infty} dk_0 \frac{e^{-ik_0 z_0}}{(k_0^2 - \kappa^2)^2}. \tag{7.21}$$

The integrand has two poles of order 2 at $k_0 = \pm\kappa$. Invariant Green's functions $D(z)$ are obtained by choosing different contours of integration relative to the two poles of order 2. Similar to the retarded Green's function with $z_0 > 0$ in electrodynamics,[5] we use Cauchy's theorem and obtain

$$\int_{-\infty}^{\infty} dk_0 \frac{e^{-ik_0 z_0}}{(k_0^2 - \kappa^2)^2} = \frac{-\pi}{\kappa^3} [sin(\kappa z_0) - z_0 \kappa cos(\kappa z_0)], \tag{7.22}$$

for $z_0 > 0$. Using (7.21) and (7.22), the invariant Green's function $D(z)$ can be expressed as

$$D(z^\lambda) = \frac{1}{(2\pi)^4} \int e^{-i\mathbf{k}\cdot\mathbf{z}} d^3k \frac{\pi}{\kappa^3}[z_0\kappa\cos(\kappa z_0) - \sin(\kappa z_0)],$$

$$= \int_0^\infty d\kappa \left[\frac{z_0\sin(\kappa z)\cos(\kappa z_0)}{(2\pi)^2\kappa z} - \frac{\sin(\kappa z)\sin(\kappa z_0)}{(2\pi)^2\kappa^2 z}\right],$$

where $z = |\mathbf{z}|$. Finally, we obtain the following 'retarded' phason Green's function[b] $D(z^\lambda)$,

$$D(z^\lambda) = \frac{1}{16\pi} - \frac{1}{8\pi} = -\frac{1}{16\pi}, \quad z_0 = z > 0, \tag{7.23}$$

$$D(z^\lambda) = 0 - \frac{1}{8\pi} = -\frac{1}{8\pi}, \quad z_0 > z > 0, \tag{7.24}$$

$$D(z^\lambda) = \frac{z_0}{8\pi z} - \frac{z_0}{8\pi z} = 0, \quad z > z_0 > 0. \tag{7.25}$$

The invariant Green's functions (7.23)–(7.25) suggest a new and intriguing picture of the massless phase field $B_\mu(x)$: Namely, once there is a source $\delta^4(z)$ in (7.16), a constant phase field instantly appears everywhere inside and on the forward light cone. It indicates that a physical phason does not propagate in space with the usual non-vanishing energy-momentum 4-vector, in sharp contrast to the physical photon.

7-3. Phasons with zero energy-momentum and Wigner's third class of particles

In the previous section, we discussed strange properties of the phase fields, as shown in (7.23)–(7.25). These are intimately related to the fourth-order phase field equations with general Yang-Mills symmetry. The most intriguing property of the quantum of the phase field, i.e., the phason, is that it appears to be a 'massless particle' with zero energy-momentum.

The best way to understand the phase field properties in (7.23)–(7.25) is to compare them with the corresponding retarded photon Green's function in (classical and quantum) electrodynamics. Since these properties are derived from a phason's invariant Green's function in vacuum in (7.16), let

---

[b]The finite value of the Green's function $D(z^\lambda)$ on the light cone $z_0 = z$ in (7.21) may not be unique. Generalized functions cannot be assigned values at isolated points.

us consider the corresponding retarded Green's function $D_r(z^\lambda)$ in electrodynamics. Namely, we consider $\partial_z^2 D_r(z) = \delta^4(z)$, $z^\mu = x^\mu - x'^\mu$. It has the solution[5]

$$D_r(z^\lambda) = \frac{\theta(z_0)}{4\pi R}\delta(z_0 - z), \qquad (7.26)$$

which vanishes for $z_0 > z > 0$ and $z > z_0 > 0$. To see a photon's propagation property, we consider the Fourier transform of $D_r(z^\lambda)$ in (7.26) with respect to z_0. We have

$$\int_{-\infty}^{\infty} e^{ik_0 z_0} D_r(z^\lambda) dz_0 = \frac{1}{4\pi R}e^{i\kappa R}, \qquad k_0 = \kappa, \qquad R = |\mathbf{R}| = |\mathbf{r} - \mathbf{r}'|, \quad (7.27)$$

which is the (photon) propagating Green's function of outgoing waves with the speed of light.[5] This suggests that the photon has the wave vector k_μ with the property $k_\mu k^\mu = k_0^2 - \kappa^2 = 0$. Consequently, a free photon satisfies the free equation $\partial^2 D_\gamma(z^\lambda) = 0$ and has the free wave solution $D_\gamma(z^\lambda) \propto exp(ik_\mu z^\mu)$, which explicitly involves the wave 4-vector k_μ with the Lorentz-Poincaré invariant relation, $k_\mu k^\mu = 0$.

Furthermore, a free photon satisfying $\partial^2 D_\gamma(x) = 0$ also has the following invariant solution[5]

$$D_\gamma(x) = \frac{1}{(2\pi)^3 i} \int e^{ik_\mu x^\mu} \epsilon(k_0)\delta(k^2) d^4 k = \frac{1}{2\pi}\epsilon(x^0)\delta(x^2). \qquad (7.28)$$

The Fourier transform of $D_\gamma(x)$ is proportional to $\epsilon(k_0)\delta(k^2)$, as shown in the second term in (7.28). It implies $k_0^2 - \mathbf{k}^2 = 0$ and $k^0 > 0$ for the energy-momentum of a photon.

It is revealing to compare the detailed properties of a phason with the properties of a free photon in (7.28), which is the solution of $\partial^2 D_\gamma(x) = 0$.

Let us determine the specific property of a free phason corresponding to that of a photon in (7.28). The free phason field B_μ satisfies the following fourth-order field equation,

$$\partial^2 \partial^2 D_B(x^2) = 0, \qquad x^2 = (x_0)^2 - \mathbf{x}^2. \qquad (7.29)$$

Gel'fand-Shilov derived the general solutions for the invariant equation $(\partial^2)^h D_B(x^2) = 0$ with $h \geq 2$ in terms of the generalized functions $(x^2 + i0)^\lambda$, which are expressed as follows,[6]

$$(\partial^2)^h (x^2 + i0)^{-2+h} = 0, \qquad h \geq 2, \qquad (7.30)$$

$$(x^2 + i0)^\lambda = (x^2)^\lambda, \quad for \quad x^2 > 0,$$

$$(x^2 + i0)^\lambda = e^{i\lambda\pi}|x^2|^\lambda, \quad for \quad x^2 < 0.$$

We note that if $(-2 + h) = \lambda$ in (7.30) is a non-negative integer, including $\lambda = 0$, the generalized functions[6] $(x^2 + i0)^\lambda$, $(x^2 - i0)^\lambda$ and $(x^2)^\lambda$ coincide. Thus, the result (7.30) leads to the unambiguous solution and its Fourier transform,

$$D_B(x^2) = (x^2 + i0)^0 = 1, \qquad h = 2,$$

$$\int e^{ip_\mu x^\mu} D_B(x^2) d^4x = (2\pi)^4 \delta^4(p). \tag{7.31}$$

The last result with $\delta^4(p) = \delta(p_0)\delta(p_1)\delta(p_2)\delta(p_3)$ implies that the phason has zero energy-momentum.

This is a surprisingly simple solution for the free phason equation (7.29) in four-dimensional space-time. We stress that Gel'fand-Shilov's systematic solutions (7.30), $(\partial^2)^h(x^2+i0)^{-2+h} = 0$ where $h \geq 2$, are important because they suggest that the simple solution $D_B(x^2)$ in (7.31) for $-2 + h = 0$ appears to be unique and that there is no other solution of the form $f(x^2)$.

In light of the previous discussions, the phason is intimately associated with the results (7.23)–(7.25), (7.29) and (7.31). They indicate that the phase field $B_\mu(x)$ does not propagate with a finite speed in space-time. In particular, (7.29) and (7.31) implies that the free phason in physical states is massless and has zero energy-momentum,

$$m = 0, \qquad p_\mu = 0, \qquad (for\ physical\ phasons). \tag{7.32}$$

This is in sharp contrast to the photon solution (7.28) on the light cone. The Fourier transform of the photon solution $D_\gamma(x)$ in (7.28) implies that a photon has the energy-momentum 4-vector $p_\mu = k_\mu$ with the property $p_0^2 - \mathbf{p}^2 = 0$ and $p_o > 0$.

These very intriguing properties of phase fields satisfying the general Yang-Mills U_1 symmetry and the fourth-order equation of the form $\partial^2\partial^2 B_\mu(x) = g_b \bar{q}\gamma_\mu q/L_b^2$ suggest new perspectives on the quark-confining fields. We will return to this topic when we consider general Yang-Mills SU_3 symmetries and color phase fields for quark confinement later.

These unusual properties of phasons appear to be completely incomprehensible if one looks at them from the conventional viewpoint of particles physics and field theories with the usual gauge symmetries. However, they are understandable and appear to be natural and necessary consequences of general Yang-Mills symmetry and Wigner's classification of particles.

Wigner classified all physical particles based on the most general space-time symmetry group — the Poincaré group. A particle is a unitary irreducible representation of the Poincaré algebra. Wigner's first class of particles have $m > 0$, his second class of particles have $m = 0$ and positive energy, and his third class of particles have zero mass and zero energy-momentum.[7] From an experimental viewpoint, physical particles are measurable packets of mass, energy and momentum. As a result, the phason with the properties in (7.32) is not qualified to be called a physical particle. However, the phason has a well-defined interaction Lagrangian and can produce confining potentials and forces between quarks, which are experimentally observable. (See Eqs. (7.36)–(7.40) and (8.15)–(8.19) below.) Furthermore, just like other physical particles, the phason is also a unitary irreducible representation of the Poincaré algebra. We shall consider the phason as a physical particle, despite its seemingly 'incomprehensible' properties and behavior.

We summarize our discussions of the phase fields and phasons as follows: *Since the phason is a quantum of field in the unitary irreducible representation of the Poincaré group, we formally treat it as a massless quantum in a quantized field such as a photon. The only difference is that at the end of calculations, we set the phasons in their physical state with the properties $p_\mu = 0$, as shown in (7.32).* Although physical phasons have the physical properties in (7.32), the phason propagator (7.2) can contribute through the intermediate states to a fermion self-mass, as shown in (7.4) and (7.13). Furthermore, the phason field gives a linear potential, as shown in (7.35)–(7.37) below. This can be pictured as due to the exchanges of a virtual phason between two quarks carrying baryonic charge, similar to the exchange of virtual photons between two charged particles in QED. A virtual photon could have the property $p_0^2 - \mathbf{p}^2 \neq 0$ and, similarly, a virtual phason could have $p_\mu \neq 0$ in the intermediate states.

7-4. Instantaneous baryonic forces between matter and antimatter half-universes

Let us consider the 'U_1 phase field' and its linear potential based on general Yang-Mills dynamic symmetry associated with baryonic charge and with the fourth-order phase field equations. The general U_1 invariant Lagrangian L_{qU_1} for quarks with baryonic charges g_b is given by[8]

$$L_{qU_1} = \frac{L_s^2}{2}\left(\partial^\mu B_{\mu\lambda}\partial_\nu B^{\nu\lambda}\right) + i\overline{q}(x)\gamma^\mu(\partial_\mu + ig_b B_\mu)q(x) - m_q\overline{q}(x)q(x), \quad (7.33)$$

$$\partial^2\partial^\mu B_{\mu\nu} = (g_b/L_s^2)\overline{q}\gamma_\nu q, \qquad B_{\mu\nu} = \partial_\mu B_\nu - \partial_\nu B_\mu. \quad (7.34)$$

The summation in the terms involving quarks, e.g., $m_q\overline{q}(x)q(x)$ is understood. All quarks are treated as singlets in general U_1 symmetry. Each quark in (7.33) plays a role similar to electrons in QED and carries a baryon charge g_ℓ. Based on universal gauge symmetry, we interpret L_s in (7.33) as a characteristic length of a physical system associated with general Yang-Mills symmetry. Although such an interpretation may seem strange, it is consistent with the spirit of unification.

We assume that the general U_1 phase field in (7.33) can be quantized using path integrals just like the usual gauge fields, so that we can calculate the vacuum-to-vacuum amplitudes[4] and obtain the Feynman-Dyson rules for Feynman diagrams, similar to QED.

Following the steps from (3.18) to (3.21) in Ch. 3, the potential energy $\overline{U}(r)$ between a quark q and an antiquark $\bar{q}$ is

$$\overline{U}(r) = (-g_s)B_0(r) = +\frac{g_b^2}{L_s^2}\frac{r}{8\pi}. \tag{7.35}$$

The corresponding attractive baryonic force $F_{q\bar{q}}$ is given by

$$\mathbf{F}_{q\bar{q}} = -\boldsymbol{\nabla}\overline{U}(r) = -\frac{1}{8\pi}\frac{g_b^2}{L_s^2}\frac{\mathbf{r}}{r}, \tag{7.36}$$

for two point-like baryon charges.

We can also calculate the baryonic force between, say, two gigantic spheres of uniform baryonic charge, representing the two half-universes of the Big Jets model. Each sphere has a radius R_o and mass M_{hu}.[9] The dominant term turns out to be a distance-independent force, i.e., the Okubo attractive force, between two spheres with opposite baryon charges,

$$\mathbf{F}^{Ok} \approx -\left(\frac{(3g_b)^2 M_{hu}^2}{8\pi L_s^2 m_p^2}\right)\frac{\mathbf{r}}{r}, \qquad r > 2R_o, \tag{7.37}$$

where m_p is the proton mass and each baryon (or proton) has three quarks and the baryon charge $3g_b$. Note that the Okubo force (7.37) holds under the condition $r > 2R_o$. (See Appendix 1 in Ch. 3.)

Result (7.37) leads to a corresponding linear Okubo potential $U^{Ok}(r)$ between the two matter and antimatter half-universe spheres due to baryonic charges,

$$U^{Ok}(r) = \left(\frac{(3g_b)^2 M_{hu}^2}{8\pi L_s^2 m_p^2}\right)r, \qquad r > 2R_o. \tag{7.38}$$

The matter (or baryon) half-universe and the antimatter (or antibaryon) half-universe could be permanently bound together by such a linear Okubo potential (7.38), similar to the permanent quark confinement by an empirical linear potential for charmonium as discussed by the Cornell group.[10]

What is the nature of the propagation of the forces between two such gigantic objects separated by a super-macroscopic distance? If the baryonic force (7.37) were to propagate with the speed of light, like electromagnetic waves in vacuum, it appears too 'farfetched' to consider one object as exerting a force that influences the motion of another object, if it would take more than 10 billion years to produce and observe the effects. It is gratifying that this is not the case for baryon dynamics. The baryonic force of a proposed matter half-universe would exert a force on the antimatter half-universe instantaneously, just like the instantaneous Coulomb force. In QED, the Coulomb force is due to the exchange of a time-like photon and a longitudinal photon together.[11] In the case of the baryonic force, the physical reason is related to the property that baryonic phase fields $B_\mu(x)$ satisfy a fourth-order differential equation.[12] Once the source $\delta^4(x)$ appears in the phase equation (7.16), a constant phase field instantly appears everywhere inside and on the forward light cone at time $x^0 = 0$, as shown in Eqs. (7.23)–(7.25).

References

1. J. D. Bjorken and S. D. Drell, *Relativistic Quantum Fields* (McGraw-Hill, New York, 1965), pp. 381–390.
2. J. P. Hsu and L. Hsu, *Space-Time, Yang-Mills Gravity, and Dynamics of Cosmic Expansion* (World Scientific, 2020), pp. 161–163, pp. 167–169 (for dimensional regularization), pp. 266–269. For a review of regularizations, see pp. 103–109 in Muta's book, *Foundations of Quantum Chromodynamics*. (World Scientific).
3. J. J. Sakurai, *Advanced Quantum Mechanics* (Addison-Wesley, 1967) pp. 268–270, pp. 273–275.
4. J. P. Hsu and L. Hsu, Ref. 2, p. 149 for vacuum-to-vacuum amplitude with an external source.
5. J. D. Jackson, *Classical Electrodynamics* (2nd ed. John Wiley & Sons, New York, 1999) pp. 223–225, pp. 608–611. N. N. Bogoliubov and D. V. Shirkov, *Introduction to the Theory of Quantized Fields* (Interscience Publishers, 1959) 648–652.
6. I. M. Gel'fand, G. E. Shilov, *Generalized Functions*, vol. 1, (New York: Academic Press, 1964). p. 280, p. 363. See also Ref. 12.
7. E. Wigner, Ann. Math. **40** 149 (1939).
8. J. P. Hsu, Mod. Phys. Letters **A** 1450031 (2014).
9. J. P. Hsu, L. Hsu and D. Katz, Mod. Phys. Letters, A **33** 1850116 (2018).
10. E. Eichten, K. Gottfried, T. Kinoshita, J. Kogut, K. D. Lane, T.-Y. Yan, Phys. Rev. Lett., **34**: 369 (1975).
11. J. J. Sakurai, Ref. 3, p. 255.

12. It may be interesting to compare the solutions of the Lorentz-Poincaré invariant equations

$$(\partial^2)^n D(x^2) = 0, \quad n = 1, 2, 3; \quad x^2 = (x_0)^2 - \mathbf{x}^2,$$

and their Fourier transforms (FT)

$$\partial^2 D_\gamma = 0, \quad D_\gamma \propto \epsilon(x_0)\delta(x^2), \quad FT: \propto \epsilon(p_0)\delta(p^2); \quad (photon),$$

where $p^2 = (p_0)^2 - \mathbf{p}^2$. Using Gel'fand-Shilov solutions (7.30), one has

$$\partial^2 \partial^2 D_B = 0, \quad D_B \propto 1, \quad FT: \propto \delta^4(p); \quad (phason)$$

$$\partial^2 \partial^2 \partial^2 D_U = 0, \quad D_U \propto (x^2 + i0), \quad FT: \propto (p^2 - i0)^{-3};$$

where[6]

$$(x^2 + i0)^\lambda = (x^2)^\lambda, \quad for \quad x^2 > 0,$$

$$(x^2 + i0)^\lambda = e^{i\lambda\pi}|x^2|^\lambda, \quad for \quad x^2 < 0.$$

8

Quark Confinement and the Accelerated Cosmic Expansion

8-1. A symmetry-unified quark-cosmic model

The physics of interactions that govern phenomena at the smallest scales (e.g., the strong force responsible for quark confinement) and at the largest scales (e.g., the repulsive baryon-baryon force possibly responsible for the late-time accelerated cosmic expansion) can both be described within a new $SU_3 \times U_1$ framework in general Yang-Mills symmetry. Although these two different physical phenomena are separately related to particle physics and cosmology, we will demonstrate in this chapter that they can be 'unified' on the basis of general Yang-Mills symmetry. Such a unification could be termed a 'symmetry-unification' in the sense that they are dictated by the same general Yang-Mills symmetry and the same universal length L_s in the Lagrangian. Although the two interactions do not seem to have the same coupling constant, both forces originate from a linear potential generated from two different types of conserved charges associated with the same general Yang-Mills symmetry, i.e., the color charges of quarks and the baryonic charge.

The conventional usage of the term 'particle-cosmology' refers to a framework based on particle physics and Einstein gravity, it is not a consistent framework. Particle physics and quantum field theories are based on a flat space-time with the Poincaré group in inertial frames. However, Einstein gravity is based on the principle of general coordinate invariance in curved space-time, which is incompatible with the Poincaré group and does not have a reference frame with operationally defined space-time coordinates.[1] Mathematically, space-time coordinates in the Riemann manifold (or curved space-time) can only be defined locally and do not have meaning by themselves, as stressed by S. S. Chern.[2,3]

In contrast, the present discussion is based on Broader Particle-Cosmology, which includes particle physics and (quantum and classical) Yang-Mills gravity in flat space-time, as discussed in Ch. 1. It also includes general frames of reference, including both inertial and non-inertial frames, whose space-time coordinates are defined operationally.

Based on Broader Particle-Cosmology, we now consider a 'quark-cosmic' model with the internal symmetry groups, $(SU_3)_{color} \times (U_1)_{baryon}$, in the absence of gravity for simplicity. The effects of very small violations of all internal gauge symmetries by Yang-Mills gravity will be discussed in Ch. 10. Such a $SU_{3c} \times U_{1b}$ model is consistent with all well-established conservation laws, which include the conservation of color charge and the conservation of baryon charge (or number).[4,5] The novel symmetries are based on general Yang-Mills (gYM) transformations, in which the usual scalar gauge functions ω^a and phase factors are replaced by the Lorentz vector gauge functions ω^a_μ and Hamilton's characteristic phase factors, as discussed in Ch. 6.

For physics in the smallest quark world, a burning question is: What is the explicit mechanism for quark confinement? The present Broader Particle-Cosmology with gYM symmetry leads to fourth-order equations for massless phase fields. In the static limit, the new phase equations could lead to dual linear and Coulomb-type potentials, together with a universal length L_s associated with the gYM invariant Lagrangian. Thus, gYM symmetry provides an explicit mechanism to confine quarks. The quark confining field is denoted by $C^a_\mu(x)$ and we call its quantum a 'confion' (C).

Fourth-order field equations are usually considered unphysical because the associated dynamical systems then have a non-definite energy (i.e., gauge bosons associated with fourth-order field equations have negative energy). However, in the present framework of Broader Particle-Cosmology, the quanta of phase fields, i.e., 'phasons,' are confined and do not propagate. Therefore, they are not directly observable and hence, do not contradict experiments.

For physics in the largest cosmic world, a burning question is: Do we have a field-theoretic understanding of the accelerated cosmic expansion? Based on Broader Particle-Cosmology, the answer turns out to be the affirmative, because all (internal) conserved quantities are associated with general Yang-Mills symmetries. To be specific, there is a gYM U_1 phase field for the baryon charge (although such a U_1 phase field has not been detected in high energy laboratories, presumably due to its extremely weak coupling strength). We argued in Sec. 4 of Ch. 3 that this is probably the case and that such an extremely weak force might be evident only at cosmic scales where an enormous number of baryons are involved. This force is associated with gYM invariant fourth-order field equations, which lead to a static linear potential between baryonic charges. Only when the universe has been expanding for long enough will the distances between baryonic galaxies be large enough for the super-weak baryonic repulsive force to overcome the gravitational attractive force between them.

This property appears to be precisely what experimentalists have observed regarding the late-time accelerated cosmic expansion. Before a critical distance between galaxies R_c is reached, the attractive gravitational force will dominate and one should observe a decelerated, rather than an accelerated cosmic expansion. As discussed in Ch. 3, this is a prediction of the present symmetry-unified particle-cosmology model. Thus, instead of the conventional assumption of an ad hoc dark energy, the U_1 phase fields could provide an understanding of the late-time accelerated cosmic expansion based on the universal principle of general Yang-Mills symmetry.

8-2. General Yang-Mills transformations for $SU_3 \times U_1$ groups and the invariant Lagrangian with higher-order derivatives

To discuss the general Yang-Mills (gYM) invariant Lagrangian and the corresponding Feynman-Dyson rules in the quark-cosmic model, let us briefly review the new gYM transformations and their consequences. The general $SU_3 \times U_1$ transformations for baryons and quarks $q(x)$, which carry color charge, are

$$q' = (1 - iP_3 - iP_1)q, \quad \bar{q}' = \bar{q}(1 + iP_3 + iP_1), \tag{8.1}$$

$$P_3 = g_s \left(\int^x dx'^\mu \omega_\mu^a(x') \frac{\lambda^a}{2} \right)_{Le3}, \quad P_1 = g_b \left(\int^x dx'^\mu \Lambda_\mu(x') \right)_{Le1}, \tag{8.2}$$

where P_3 and P_1 are Hamilton's characteristic functions, and ω_μ^a and Λ_μ are arbitrary infinitesimal (Lorentz) vector functions. Moreover, the gYM transformations for the color SU_3 confion fields $C_\mu^a(x)$ and the U_1 phase field $B_\mu(x)$, are given by

$$B'_\mu(x) = B_\mu(x) + \Lambda_\mu(x), \tag{8.3}$$

$$C'_\mu(x) = C_\mu(x) + \omega_\mu(x) - i[P_3, C_\mu], \tag{8.4}$$

$$C_\mu = C_\mu^a \frac{\lambda^a}{2}, \quad \left[\frac{\lambda^a}{2}, \frac{\lambda^b}{2} \right] = if^{abc}\frac{\lambda^c}{2}, \quad a,b,c = 1,...8,$$

where C_μ^a is the SU_3 confion (or phase) field, B_μ is the U_1 baryonic phase field, and f^{abc} is real and completely antisymmetric in a, b, c.[6] The gauge vector function Λ_μ in the gYM transformation (8.2) satisfies four constraints, as shown in (8.11) below. The quark transformation in (8.1) involves a Hamilton's characteristic phase factor, as explained in Ch. 6.

The form of the gYM gauge covariant derivatives are the same as those of the usual internal gauge groups. As usual, the $SU_3 \times U_{1b}$ gauge covariant derivatives are defined as follows,

$$\Delta_\mu = \partial_\mu - ig_s C_\mu^a \frac{\lambda^a}{2} - ig_b B_\mu. \qquad (8.5)$$

The SU_3 and U_{1b} gauge curvatures $C_{\mu\nu}^a$ and $B_{\mu\nu}$ are given by[7]

$$[\Delta_\mu, \Delta_\nu] = -ig_s C_{\mu\nu} - ig_b B_{\mu\nu}, \qquad (8.6)$$

$$C_{\mu\nu} = \partial_\mu C_\nu - \partial_\nu C_\mu - ig_s[C_\mu, C_\nu], \quad B_{\mu\nu} = \partial_\mu B_\nu - \partial_\nu B_\mu. \qquad (8.7)$$

Following from equations (8.1)-(8.4), we now have the general Yang-Mills transformations for $\partial^\mu C_{\mu\nu}(x)$, $\partial^\mu B_{\mu\nu}(x)$ and $\bar{q}\Delta_\mu q$:

$$\partial^\mu B'_{\mu\nu}(x) = \partial^\mu B_{\mu\nu}(x) + \partial^\mu \partial_\mu \Lambda_\nu(x) - \partial^\mu \partial_\nu \Lambda_\mu(x) = \partial^\mu B_{\mu\nu}(x), \qquad (8.8)$$

$$\partial^\mu C'_{\mu\nu}(x) = \partial^\mu C_{\mu\nu}(x) - [P_3, \partial^\mu C_{\mu\nu}], \qquad (8.9)$$

$$\bar{q}'\gamma^\mu \Delta'_\mu q' = \bar{q}\gamma^\mu \Delta_\mu q, \qquad (8.10)$$

provided the following restrictions for the U_1 and SU_3 vector gauge functions Λ_μ and ω_μ^a are imposed,

$$\partial^2 \Lambda_\mu(x) - \partial_\mu \partial^\lambda \Lambda_\lambda(x) = 0, \qquad (8.11)$$

$$\partial^\mu \{\partial_\mu \omega_\nu(x) - \partial_\nu \omega_\mu(x)\} - ig_s[\omega^\mu(x), C_{\mu\nu}(x)] = 0. \qquad (8.12)$$

We have also used the relations,

$$\partial_\mu P_1 = g_b \Lambda_\mu(x), \quad and \quad \partial_\mu P_3 = g_s \omega_\mu^a(x) \frac{\lambda^a}{2}, \qquad (8.13)$$

where P_1 and P_3 are Hamilton's characteristic functions as defined in (8.2). The space-time derivative of the gauge curvatures have the desired transformation properties, as shown in (8.9).

8-3.   A novel quark-confining mechanism

Let us now concentrate on the static potentials in the SU_3 sector by setting $g_b = 0$. The static potential produced by the general U_1 baryonic phase field B_μ was discussed in Ch. 3. The general SU_3 invariant Lagrangian L_{su_3} takes the form

$$L_{su_3} = \frac{L_s^2}{2}[\partial^\mu C_{\mu\lambda}^a \partial_\nu C^{a\nu\lambda}] + i\bar{q}[\gamma^\mu \Delta_\mu - M]q, \qquad (8.14)$$

$$C^a_{\mu\nu} = \partial_\mu C^a_\nu - \partial_\nu C^a_\mu + g_s f^{abc} C^b_\mu C^c_\nu, \qquad \Delta_\mu = \partial_\mu + i g_s C^a_\mu \frac{\lambda^a}{2}.$$

The matter fields consists of spinor quark fields $q(x)$ with components q^{fi}, where $i = 1, 2, 3$ for color indices and $f = 1, 2, ...6$ for flavor indices (u,d,c,s,t,b), and M is a color-independent mass matrix in the flavor indices.[8] The constant L_s denotes a length scale that characterizes the dynamical systems described by the gYM invariant Lagrangian (8.14). We postulate that L_s is a universal characteristic length for phase fields with general Yang-Mills symmetry and is an essential feature of unification — a feature common to both the tiny quark world and the gigantic cosmic world.

From the Lagrangian (8.14) with general Yang-Mills symmetry, one can derive the fourth-order field equation for the confion fields C^a_μ,

$$\partial^2 \partial^\mu C^a_{\mu\nu} + (g_s/L_s^2)\bar{q}\gamma_\nu(\lambda^a/2)q = 0. \tag{8.15}$$

To simplify the confion field equation, we impose a gauge condition $\partial^\mu C^a_\mu = 0$, obtaining the confion equation

$$\partial^2 \partial^2 C^a_\mu = -(g_s/L_s^2)\bar{q}\gamma_\nu(\lambda^a/2)q - g_s \partial^2 [\partial^\mu(f^{abc} C^b_\mu C^c_\nu)] + \dots. \tag{8.16}$$

In the static case, there are two types of singular sources: The first term in (8.16) is the usual quark source, which corresponds to the usual static source $-g_s \delta^3(\mathbf{r})$. The second term involves $\partial^2 = \partial_t^2 - \nabla^2$ acting on a usual source function and is interpreted to correspond to a new type of singular source $-L_s^2 \nabla^2 \delta^3(\mathbf{r})$ in the static limit. This new source could be generated from the self-coupling of the confion phase fields. Thus, for a given index a, we have the following roughly approximate static equation and the confion potential $C_0 = C_0(r)$,

$$\nabla^2 \nabla^2 C_0 = -\frac{g_s}{L_s^2}\delta^3(\mathbf{r}) + g_s \nabla^2 \delta^3(\mathbf{r}), \tag{8.17}$$

$$C_0 = g_s \left[\frac{r}{8\pi L_s^2} - \frac{1}{4\pi r} \right]. \tag{8.18}$$

If we focus on charmed quarks in (8.15), we can derive the confining potential energy $V = g_s C_0$ for charmonium and estimate the associated color charge g_s and scale length L_s as roughly[8]

$$\frac{g_s^2}{4\pi} \approx 0.2, \qquad L_s \approx 0.14 fm, \tag{8.19}$$

where the mass of the charm quark is assumed to be $m_c \approx 1.6 GeV$. The key feature of this confining mechanism is that the confion is governed by the fourth-order field equations, which have two types of singular sources that produce dual linear and Coulomb-like potential. Therefore, the confining quark potentials (8.18) and (8.19) are just right to support the ideas and results of the Cornell group — Eichten, Gottfried, Kinoshita, Kogut, Lane and Yan.[8]

8-4. Feynman-Dyson rules for the confining quark model

Let us consider a confining quark model with confions, which generate the dual confining potentials in (8.17). To derive rules for the Feynman diagrams, we include as usual a gauge fixing term for $C_\mu^a(x)$ in the Lagrangian. The confion Lagrangian L_{su_3} with gauge fixing terms involving the gauge parameter η is employed to quantize the fields as usual.

Since we are interested in the Feynman-Dyson rules for the confining quark model, let us consider the following total Lagrangian L_{tot} involving confion fields $C_\mu^a(x)$ and quarks,

$$L_{tot} = L_{su_3 u_1} + L_{gf}, \qquad (8.20)$$

$$L_{su_3 u_1} = \frac{L_s^2}{2} \left[\partial^\mu C_{\mu\lambda}^a \partial_\nu C^{a\nu\lambda} + \partial^\mu B_{\mu\lambda} \partial_\nu B^{\nu\lambda} \right] \qquad (8.21)$$

$$+ i\bar{q} \left[\gamma^\mu (\partial_\mu - ig_s C_\mu^a \frac{\lambda^a}{2} - ig_b B_\mu) - M \right] q$$

$$L_{gf} = +L_s^2 \left[\frac{1}{2\eta} \partial_\lambda \partial_\mu C^{a\mu} \partial^\lambda \partial_\nu C^{a\nu} + \frac{1}{2\eta'} \partial_\lambda \partial_\mu B^\mu \partial^\lambda \partial_\nu B^\nu \right]. \qquad (8.22)$$

The field equations for C_μ^a and B_μ can be derived from the Lagrangian L_{tot} in (8.20),

$$\partial^2 \partial^\mu C_{\mu\nu}^a - \left(1 - \frac{1}{\eta} \right) \partial^2 \partial_\mu \partial_\nu C^{a\nu} = \frac{g_s}{L_s^2} \bar{q} \gamma_\mu \frac{\lambda^a}{2} q, \qquad (8.23)$$

$$\partial^2 \partial^\mu B_{\mu\nu} - \left(1 - \frac{1}{\eta'} \right) \partial^2 \partial_\mu \partial_\nu B^\nu = \frac{g_b}{L_s^2} \bar{q} \gamma_\mu q.$$

These lead to the confion propagator $c_{\mu\nu}^{ab}(k)$ and the baryonic phason propagator $b_{\mu\nu}(k)$ which, in the Feynman gauge $\eta = \eta' = 1$, are given by

$$c_{\mu\nu}^{ab}(k) = \frac{-i\delta^{ab}\eta_{\mu\nu}}{L_s^2 (k^2 + i\epsilon)^2}, \qquad b_{\mu\nu}(k) = \frac{-i\eta_{\mu\nu}}{L_s^2 (k^2 + i\epsilon)^2}. \qquad (8.24)$$

A physical particle (i.e., quantum) in field theory is a unitary irreducible representation of the Poincaré group, characterized by a spin and a mass

$m \geq 0$. Wigner divided particles into three classes. The first class consists of particles with $m > 0$. The second class consists of particles with $m = 0$ and positive energy. Particles from both these classes have been observed in the laboratory. Wigner also named a third class of particles with zero mass and zero energy-momentum, as discussed in Ch. 7. We propose that confions, if they exist, are an example of this third class of particles.

As shown in (8.23), confions satisfy fourth-order field equations and general Yang-Mills symmetry. According to the previous discussion in Sec. 7-3, they are massless particles with zero energy-momentum. They do not propagate in vacuum and cannot be directly detected. Although they cannot be directly observed as free physical particles in external states, they can be exchanged between quarks in the intermediate states of a physical process and produce confining potential between quarks, as shown in (8.17) and (8.18). Thus, a test of this framework of Broader Particle-Cosmology can also provide an indirect test of the existence of Wigner's third class of particles.

To elaborate the field-theoretic picture of a virtual confion exchange related to the solution to (8.17) and (8.18), let us use the method of Fourier transforms involving generalized functions.[9] We define

$$C_0(r) = \int e^{i\mathbf{k}\cdot\mathbf{r}} C_0(k) \frac{d^3k}{(2\pi)^3}, \qquad k = |\mathbf{k}|, \tag{8.25}$$

$$C_0(k) = \int e^{-i\mathbf{k}\cdot\mathbf{r}} C_0(r) d^3r. \tag{8.26}$$

From (8.17) and (8.25), we have

$$\nabla^2 \nabla^2 C_0(r) = \int (k^2)^2 e^{i\mathbf{k}\cdot\mathbf{r}} C_0(k) \frac{d^3k}{(2\pi)^3} \tag{8.27}$$

$$= \int \left[\frac{-g_s}{L_s^2} - g_s k^2 \right] e^{i\mathbf{k}\cdot\mathbf{r}} \frac{d^3k}{(2\pi)^3},$$

which gives

$$C_0(k) = \frac{1}{(k^2)^2} \left[\frac{-g_s}{L_s^2} - g_s k^2 \right]. \tag{8.28}$$

From (8.25) and (8.28), we obtain dual linear and Coulomb-like potentials, $C_0(r)$,

$$C_0(r) = \int e^{i\mathbf{k}\cdot\mathbf{r}} \frac{1}{k^2 k^2} \left[\frac{-g_s}{L_s^2} - g_s k^2 \right] \frac{d^3k}{(2\pi)^3} = \frac{g_s r}{8\pi L_s^2} - \frac{g_s}{4\pi r}, \tag{8.29}$$

where we have used the Fourier transform of generalized functions,[9]

$$\int k^\lambda e^{i\mathbf{k}\cdot\mathbf{r}} d^3 k = 2^{\lambda+3} \pi^{3/2} \frac{\Gamma([\lambda+3]/2)}{\Gamma(-\lambda/2)} r^{-\lambda-3}, \qquad (8.30)$$

with $\Gamma(-1/2) = -2\sqrt{\pi}$ and $\Gamma(1/2) = \sqrt{\pi}$.

Since the confions are permanently confined inside the quark system, we approximate these two static sources as having the same position, as shown in (8.29).

It is natural to treat confions formally in a similar way to other quanta such as photons in the formulation of the S matrix or the Feynman-Dyson rules, since they are unitary irreducible representations of the Poincaré group. They appear in the intermediate states of a physical process and contribute to observable results (e.g., the confining potentials). However, they do not appear in the external particle states of the S matrix and, when they do appear in intermediate states, they do not contribute to the imaginary parts of amplitudes and upset the unitarity of the S matrix. Similar to photons in QED, confions do not need a ghost associated with a functional determinant in the vacuum-to-vacuum amplitude of the confining quark model.

Thus, the vacuum-to-vacuum amplitude $W(y_\lambda^a)$ for the SU_3 sector is assumed to be

$$W(y_\lambda^a) = \int d[C, q, \bar{q}] exp\left(i \int L_{su3} d^4 x\right) \delta(\partial_\lambda \partial_\mu C^{\mu a} - y_\lambda^a). \qquad (8.31)$$

Based on the Lagrangian L_{tot} in (8.20), the confion propagator $c_{\mu\nu}^{ab}(k)$ with a general gauge parameter η is given by

$$c_{\mu\nu}^{ab}(k) = \frac{-i\delta^{ab}}{L_s^2(k^2 + i\epsilon)^2}\left[\eta_{\mu\nu} - (1 - \eta)\frac{k_\mu k_\nu}{k^2 + i\epsilon}\right], \qquad (8.32)$$

which reduces to (8.24) in the Feynman gauge, $\eta = 1$. The quark propagator is given by

$$\frac{i\delta_{jk}\delta_{mn}}{p_\mu\gamma^\mu - M}, \qquad (8.33)$$

where (j, k) are color indices and (m, n) are flavor indices. For the confion-quark 3-vertex and the baryonic 3-vertex (qqB_μ), we have

$$ig_s\gamma^\mu\frac{\lambda^a}{2} \quad and \quad ig_b\gamma^\mu \qquad (8.34)$$

respectively.

The 3-confion vertex $[C_\alpha^a(k_1)C_\beta^b(k_2)C_\gamma^c(k_3)]$ can be obtained from the Lagrangian L_{su3},

$$g_s L_s^2 f^{abc}[(k_1)^2(k_{2\beta}\eta_{\alpha\gamma} - k_{3\gamma}\eta_{\alpha\beta} + k_{3\beta}\eta_{\alpha\gamma} - k_{2\gamma}\eta_{\alpha\beta})$$

$$+(k_2)^2(k_{3\gamma}\eta_{\alpha\beta} - k_{1\alpha}\eta_{\gamma\beta} + k_{1\gamma}\eta_{\alpha\beta} - k_{3\alpha}\eta_{\gamma\beta})$$

$$+(k_3)^2(k_{1\alpha}\eta_{\beta\gamma} - k_{2\beta}\eta_{\alpha\gamma} + k_{2\alpha}\eta_{\beta\gamma} - k_{1\beta}\eta_{\alpha\gamma})], \tag{8.35}$$

where $(k_n)^2 = k_{n\mu}k_n^\mu$, $n = 1,2,3$.

The 4-confion vertex $[C_\alpha^a(k_1)C_\beta^b(k_2)C_\gamma^c(k_3)C_\delta^d(k_4)]$ is given by

$$-\frac{i}{2}L_s^2 g_s^2[f^{eab}f^{ecd}k_{1\alpha}(k_{3\gamma}\eta_{\delta\beta} - k_{4\delta}\eta_{\beta\gamma}) + f^{eac}f^{ebd}k_{1\alpha}(k_{2\beta}\eta_{\gamma\delta} - k_{4\delta}\eta_{\beta\gamma})$$

$$+f^{ead}f^{ebc}k_{1\alpha}(k_{2\beta}\eta_{\delta\gamma} - k_{3\gamma}\eta_{\beta\delta}) + f^{eba}f^{ecd}k_{2\beta}(k_{3\gamma}\eta_{\alpha\delta} - k_{4\delta}\eta_{\alpha\gamma})$$

$$+f^{ebc}f^{ead}k_{2\beta}(k_{1\alpha}\eta_{\delta\gamma} - k_{4\delta}\eta_{\alpha\gamma}) + f^{ebd}f^{eac}k_{2\beta}(k_{1\alpha}\eta_{\delta\gamma} - k_{3\gamma}\eta_{\alpha\delta})$$

$$+f^{eca}f^{ebd}k_{3\gamma}(k_{2\beta}\eta_{\delta\alpha} - k_{4\delta}\eta_{\beta\gamma}) + f^{ecb}f^{ead}k_{3\gamma}(k_{1\alpha}\eta_{\delta\beta} - k_{4\delta}\eta_{\beta\alpha})$$

$$+f^{ecd}f^{eab}k_{3\gamma}(k_{1\alpha}\eta_{\delta\beta} - k_{2\beta}\eta_{\delta\alpha}) + f^{eda}f^{ebc}k_{4\delta}(k_{2\beta}\eta_{\alpha\gamma} - k_{3\gamma}\eta_{\beta\alpha})$$

$$+f^{edb}f^{eac}k_{4\delta}(k_{1\alpha}\eta_{\gamma\beta} - k_{3\gamma}\eta_{\beta\alpha}) + f^{edc}f^{eab}k_{4\delta}(k_{1\alpha}\eta_{\gamma\beta} - k_{2\beta}\eta_{\alpha\gamma})$$

$$-iL_s^2 g_s^2[f^{eab}f^{ecd}k_{1\alpha}(k_{4\gamma}\eta_{\delta\beta} - k_{3\delta}\eta_{\beta\gamma}) + f^{eac}f^{ebd}k_{1\alpha}(k_{4\beta}\eta_{\delta\gamma} - k_{2\delta}\eta_{\beta\gamma})$$

$$+f^{ead}f^{ebc}k_{1\alpha}(k_{3\beta}\eta_{\delta\gamma} - k_{2\gamma}\eta_{\beta\delta}) + f^{eba}f^{ecd}k_{2\beta}(k_{4\gamma}\eta_{\delta\alpha} - k_{3\delta}\eta_{\alpha\gamma})$$

$$+f^{ebc}f^{ead}k_{2\beta}(k_{4\alpha}\eta_{\delta\gamma} - k_{1\delta}\eta_{\alpha\gamma}) + f^{ebd}f^{eac}k_{2\beta}(k_{3\alpha}\eta_{\delta\gamma} - k_{1\gamma}\eta_{\delta\alpha})$$

$$+f^{eca}f^{ebd}k_{3\gamma}(k_{3\beta}\eta_{\delta\alpha} - k_{2\delta}\eta_{\beta\alpha}) + f^{ecb}f^{ead}k_{3\gamma}(k_{4\alpha}\eta_{\delta\beta} - k_{1\delta}\eta_{\beta\alpha})$$

$$+f^{ecd}f^{eab}k_{3\gamma}(k_{2\alpha}\eta_{\delta\beta} - k_{1\beta}\eta_{\alpha\delta}) + f^{eda}f^{ebc}k_{4\delta}(k_{3\beta}\eta_{\alpha\gamma} - k_{2\gamma}\eta_{\beta\alpha})$$

$$+f^{edb}f^{eac}k_{4\delta}(k_{3\alpha}\eta_{\gamma\beta} - k_{1\gamma}\eta_{\beta\alpha}) + f^{edc}f^{eab}k_{4\delta}(k_{2\alpha}\eta_{\gamma\beta} - k_{1\beta}\eta_{\alpha\gamma})$$

$$-\frac{i}{2}L_s^2 g_s^2[f^{eab}f^{ecd}k_{2\alpha}(k_{4\gamma}\eta_{\delta\beta} - k_{3\delta}\eta_{\beta\gamma}) + f^{eac}f^{ebd}k_{3\alpha}(k_{4\beta}\eta_{\delta\gamma} - k_{2\delta}\eta_{\beta\gamma})$$

$$+f^{ead}f^{ebc}k_{4\alpha}(k_{3\beta}\eta_{\delta\gamma} - k_{2\gamma}\eta_{\beta\delta}) + f^{eba}f^{ecd}k_{1\beta}(k_{4\gamma}\eta_{\delta\alpha} - k_{3\delta}\eta_{\alpha\gamma})$$

$$+f^{ebc}f^{ead}k_{3\beta}(k_{4\alpha}\eta_{\delta\gamma} - k_{1\delta}\eta_{\alpha\gamma}) + f^{ebd}f^{eac}k_{4\beta}(k_{3\alpha}\eta_{\delta\gamma} - k_{1\gamma}\eta_{\delta\alpha})$$

$$+f^{eca}f^{ebd}k_{1\gamma}(k_{4\beta}\eta_{\delta\alpha} - k_{2\delta}\eta_{\beta\alpha}) + f^{ecb}f^{ead}k_{2\gamma}(k_{4\alpha}\eta_{\delta\beta} - k_{1\delta}\eta_{\beta\alpha})$$

$$+f^{ecd}f^{eab}k_{4\gamma}(k_{2\alpha}\eta_{\delta\beta} - k_{1\beta}\eta_{\delta\alpha}) + f^{eda}f^{ebc}k_{1\delta}(k_{3\beta}\eta_{\alpha\gamma} - k_{2\gamma}\eta_{\beta\alpha})$$

$$+f^{edb}f^{eac}k_{2\delta}(k_{3\alpha}\eta_{\gamma\beta} - k_{1\gamma}\eta_{\beta\alpha}) + f^{edc}f^{eab}k_{3\delta}(k_{2\alpha}\eta_{\gamma\beta} - k_{1\beta}\eta_{\alpha\gamma})]. \tag{8.36}$$

There is a factor -1 for each quark loop and the quark propagator in the U_1 sector has the same form as the electron propagator in QED.[10]

References

1. E. P. Wigner, *Symmetries and Reflections, Scientific Essays*, (The MIT Press, 1967), pp. 52–53 (in collaboration with H. Salecker). Wigner wrote: 'Evidently, the usual statements about future positions of particles, as specified by their coordinates, are not meaningful statements in general relativity. This is a point which cannot be emphasized strongly enough ....'
2. S. S. Chern (with W. H. Chen), *Lecture Notes on Differential Geometry*, (Lian Jing Publisher, Taipei, 1987) p. iv.
3. This may be a main difficulty to overcome in order to have a physically useful global differential geometry.
4. K. Huang, *Quarks, Leptons and Gauge Fields* (World Scientific, 2nd ed., 1992), pp. 105–120 and pp. 252–260.
5. T. D. Lee, *Particle Physics and Introduction to Field Theory.* (Harwood Academic Publishers, 1981). Ch. 10.
6. K. Huang, in Ref. 4, p. 17.
7. J. P. Hsu and L. Hsu, *Space-Time, Yang-Mills Gravity and Dynamics of Cosmic Expansion* (World Scientific, 2020) p. 99.
8. E. Eichten, K. Gottfried, T. Kinoshita, J. Kogut, K. D. Lane, T.-M. Yan, Phys. Rev. Lett., **34**: 369 (1975).
9. I. M. Gel'fand, G. E. Shilov, *Generalized Functions* vol. 1, (New York: Academic Press, 1964), p. 363.
10. J. D. Bjorken and S. D. Drell, *Relativistic Quantum Fields*, (McGraw-Hill, 1965), Appendix B.

9

A Simple Harmonic Oscillator Model for
3-Quark Confinement

9-1. Quark 3-body system from consecutive 2-body collisions

In this chapter, we will apply the general Yang-Mills symmetry framework to two modern problems: (A) explaining the spectrum of charmonium[1] and calculating the masses of the proton and neutron[2] from the confining potential and (B) a new model of baryon formation, which effectively reduces a complex 3-body collision problem to the much simpler problem of determining the relativistic energy of a quark in a simple harmonic oscillator potential. Such a reduction is possible only because of the special properties of the linear confining potential between two point-like color charges. This new model may help our understanding of the permanent stability of the proton and the baryon spectrum in a true test of our understanding of hadron physics.

The proton-neutron mass and stability problems have captured particle physicists' imagination for a long time.[3] Since the conception of the quark by Gell-Mann and Zweig,[4,5] particle physicists have had a more reliable framework with which to contemplate these long-standing problems. However, because calculating the masses of the proton and neutron is a 3-body problem involving the strong interaction, it has generally been considered intractable. To the best of our knowledge, there has been no calculation of the proton or neutron mass based on quantum chromodynamics.

The N resonance system $N(1440)$ has a Breit-Wigner full width $\approx 350 MeV$, and decays into a nucleon (3 quarks) and a pion (2 quarks) with an extremely short lifetime of $\approx 10^{-24}$ seconds.[2] This suggests that the formation of a quark 3-body system is extremely efficient and occurs with a high probability. Conventionally however, the probability of 3-body collisions appears to be extremely small and the possible patterns of three-body formations seem extremely complicated. Even classically, the 3-body problem is highly non-trivial and intractable. However, based on quark confining potentials, we now consider a new model that can give us a simplified picture of the formation of a 3-quark system. The basic idea of the model is that a 3-quark system such as a red-u/yellow-u/green-d proton is formed by two consecutive 2-body collisions.

To be specific, let us consider the following two interactions:

(i) A red u-quark and a yellow u-quark collide to form 'quarkonium,' a state similar to positronium, with an s-state wave function of the form $exp(-r/a_q)$, for the distribution of color charge, that produces a simple harmonic oscillator (SHO) potential.

(ii) A green d-quark, attracted by the SHO potential, collides with the quarkonium to form a stable color singlet state (in this case, a proton). The probability of this collision is greatly enhanced by the long-range strong attractive force of the SHO potential.

This model uses a specific confining SHO potential for a 3-quark system. Let us consider the force due to the color hypercharges, which correspond to an eigenvalue of $\lambda^8/2$ and can be expressed in units of g_s.[6,a] A red u-quark (with color hypercharge $g_s/(2\sqrt{3})$) and a yellow u-quark (with color hypercharge $g_s/(2\sqrt{3})$ can form a bound state similar to positronium with an s-state wave function Ψ. The hypercharge associated with this 'quarkonium' state can be modeled as a cloud shaped like a spherical shell, similar to the electronic states in the quantum mechanical atomic model.

9-2. Improved estimations of the confining potentials

Let us consider a symmetry-unified quark-cosmic model based on general Yang-Mills symmetry with the gauge groups[3] color SU_3 and baryonic U_1, i.e., $[SU_{3c} \times U_{1b}]$. The general Yang-Mills (gYM) invariant Lagrangian L_{gYM} involves confion phase fields C_μ^a and the baryonic phase field B_μ,

$$L_{gYM} = \frac{L_s^2}{2}[\partial^\mu C_{\mu\lambda}^a \partial_\nu C^{a\nu\lambda} + \partial^\mu B_{\mu\lambda}\partial_\nu B^{\nu\lambda}] \tag{9.1}$$

$$+\bar{q}[i\gamma^\mu(\partial_\mu + ig_s C_\mu^a \frac{\lambda^a}{2} + ig_b B_\mu) - M]q,$$

$$C_{\mu\lambda}^a = \partial_\mu C_\lambda^a - \partial_\lambda C_\mu^a + g_s f^{abc} C_\mu^b C_\lambda^c, \quad B_{\mu\lambda} = \partial_\mu B_\lambda - \partial_\lambda B_\mu,$$

where λ^a are eight 3×3 Gell-Mann matrices of the SU_3 group.[6] The matter fields consist of spinor quark fields $q(x)$ with components q^{fi}, where $i = 1, 2, 3$ for color indices and $f = 1, 2, ...6$ for flavor indices (u,d,c,s,t,b), and M is a color-independent mass matrix in the flavor indices.[6]

[a]According to this model, the confining force due to the color isotopic charge vanishes because the color isotopic charges for red u-quark, yellow u-quark and green u-quark are respectively $(1/2)g_s$, $(-1/2)g_s$ and 0.[6]

The length L_s denotes a universal length scale that characterizes both the dynamics of the cosmos and quark systems described by the gYM Lagrangian (9.1). It plays a dual role in Broader Particle-Cosmology as follows:

(i) It is the universal and fundamental length for all phase fields in (9.1). We will estimate its value from charmonium spectra data.[1]

(ii) Together with the baryonic charge g_b, it determines the strength of the repulsive Okubo force between two cosmic objects such as two galaxies. We estimated its value as $L_s = L_b \approx 10^{-16}$ m in eq. (3.27) in Ch. 3.

We consider first the color SU_3 sector of the gYM Lagrangian (9.1). One can derive field equations for C_μ^a and quarks,

$$L_s^2 \partial^2 \partial^\mu C_{\mu\nu}^a + g_s \bar{q} \gamma_\nu (\lambda^a/2) q = 0, \tag{9.2}$$

and

$$\left(i\gamma^\mu \left[\partial_\mu + i g_s C_\mu^a \frac{\lambda^a}{2} + i g_b B_\mu \right] - M \right) q = 0. \tag{9.3}$$

The gYM field C_μ^a satisfies the fourth-order field equation and is called the phase field, to distinguish it from the usual gauge field that satisfies the second-order field equations. The quanta of the massless phase fields C_μ^a in (9.2) are called 'confions,' whose virtual particles being exchanged between quarks produce the permanently confining potential. The sources of the confion field C_0^a include the usual quark source,

$$j_{0q}^a = -g_s \bar{q} \gamma_0 (\lambda^a/2) q, \tag{9.4}$$

and the self-coupling of 'confions', i.e.,

$$j_{0C}^a = -g_s L_s^2 f^{abc} \partial^2 [\partial^\mu (C_\mu^b C_0^c)]. \tag{9.5}$$

This term involves the operator $\partial^2 = \partial_t^2 - \nabla^2$ acting on a source function and is interpreted as a new type of singular source $-g_s L_s^2 \nabla^2 \delta^3(\mathbf{r})$ in the static limit,

$$-g_s L_s^2 \partial^2 [\partial^\mu (f^{abc} C_\mu^b C_0^c)] \to g_s L_s^2 \nabla^2 \delta^3(\mathbf{r}). \tag{9.6}$$

Thus, in the static limit, these sources for the time-component C_0 of the confion field (with a given index a) are approximated to be

$$j_0 = j_{0q} + j_{0C} \approx -g_s \delta^3(\mathbf{r}) + g_s L_s^2 \nabla^2 \delta^3(\mathbf{r}). \tag{9.7}$$

This result appears to be consistent with dimensional analysis of the source terms in (9.2). Thus, the static solution of

$$L_s^2 \nabla^2 \nabla^2 C_0 = j_0 \approx -g_s \delta^3(\mathbf{r}) + g_s L_s^2 \nabla^2 \delta^3(\mathbf{r}), \tag{9.8}$$

for each source is approximately given by

$$C_{0q} \approx g_s r/(8\pi L_s^2), \quad C_{0C} = -g_s/(4\pi r)]. \tag{9.9}$$

In this approximation, a quark is assumed to carry a unit color charge g_s. Thus, the potential energy V_q between two quarks is defined as

$$V_q = g_s C_{0q} + g_s C_{0C} \approx g_s^2 r/(8\pi L_s^2) - g_s^2/(4\pi r)], \tag{9.10}$$

so that quarks attract each other with a force $F_q = -dV_q/dr$. For the quark source $j_{0q} = -g_s \delta^3(\mathbf{r})$, the linear potential is obtained by using the Fourier transform of the generalized function,[7]

$$\int k^\lambda e^{i\mathbf{k}\cdot\mathbf{r}} d^3 k = 2^{\lambda+3} \pi^{3/2} \frac{\Gamma(-[\lambda+3]/2)}{\Gamma(-\lambda/2)} r^{-\lambda-3}, \tag{9.11}$$

where $\Gamma(-1/2) = -2\sqrt{\pi}$.

The dual quark potential energies in (9.10) are consistent with and supported by the empirical charmonium potential obtained from studying the charmonium spectrum[1] with the charm quark mass assumed to be $m_c \approx 1.6\ GeV$.

We shall use the result obtained from the charmonium spectrum to estimate g_s and L_s. The approximate confining potential energy (9.10) could be improved by considering charmonium specifically and the two specific SU_3 indices a = 8 and a = 3. In general, a quark of any color can interact with a confion of the phase fields C_μ^3 and C_μ^8 without changing color.[6] The quark source (9.4) for the charm quark $(q = q_{ci}, i = 1, 2, 3,)$ can be expressed as

$$j_{0c}^8 = -g_s \bar{q}\gamma_\nu(\lambda^8/2)q = \frac{-g_s}{2\sqrt{3}L_s^2}[\delta^3(\mathbf{r})_{11} + \delta^3(\mathbf{r})_{22} - 2\delta^3(\mathbf{r})_{33}], \tag{9.12}$$

where $\delta^3(\mathbf{r})_{kk}$ denotes the usual delta function associated with the source $\bar{q}_{ci}\gamma_0 q_{ci}$ with $i = 1, 2, 3$ denoting the color charges (say, red, yellow, green) for three different colored quarks. We have used the explicit Gell-Mann matrix[6] λ^8. Each of these three source terms in (9.12) corresponds to a specific Feynman diagram involving a color charge, which can be expressed in units of g_s. To be specific, the potential due to the exchange of a virtual confion of C_μ^8 between $\bar{q}_{ci}$ and q_{ci} satisfies the equation

$$\nabla^2\nabla^2 C_0^8 = \left(\frac{-g_s}{2\sqrt{3}L_s^2}[\delta^3(\mathbf{r})_{11} + \delta^3(\mathbf{r})_{22} - 2\delta^3(\mathbf{r})_{33}] \right. \tag{9.13}$$

$$\left. + \frac{\sqrt{3}g_s}{2}\nabla^2\delta^3(\mathbf{r}) + \frac{\sqrt{3}g_s}{2}\nabla^2\delta^3(\mathbf{r}) \right),$$

where the last two source terms are due to the self-coupling of confions, as shown in (9.5)–(9.7) with non-vanishing $f^{845} = f^{867} = \sqrt{3}/2$ for $a = 8$. The sources in the phase Eq. (9.13) give five potentials for C_0^8 and five attractive potential energies V_8. Thus, the total potential energy V_{t8} from 3 color hypercharges in the model is

$$V_{t8} = -\left[\frac{g_s^2}{12} + \frac{g_s^2}{12} + \frac{g_s^2}{3}\right]\left[\frac{r}{8\pi L_s^2}\right] + \left(\frac{3g_s^2}{4} + \frac{3g_s^2}{4}\right)\left[\frac{1}{4\pi r}\right]. \tag{9.14}$$

Similarly, for the SU_3 index $a = 3$, the confion C_0^3 between $\bar{q}_{c1}$ and q_{c1} satisfies the static equation,

$$\nabla^2\nabla^2 C_0^3 = \left(\frac{-g_s}{2L_s^2}[\delta^3(\mathbf{r})_{11} - \delta^3(\mathbf{r})_{22} - 0]\right.$$

$$\left. + g_s\nabla^2\delta^3(\mathbf{r}) + \frac{g_s}{2}\nabla^2\delta^3(\mathbf{r}) - \frac{g_s}{2}\nabla^2\delta^3(\mathbf{r})\right), \tag{9.15}$$

where we have used $f^{312} = 1$, $f^{345} = 1/2$ and $f^{367} = -1/2$ for the last three sources in (9.15).

The total attractive potential energy V_{t3} from 3 color isotopic charges in the model is

$$V_{t3} = -\left[\frac{g_s^2}{4} + \frac{g_s^2}{4} + 0\right]\left[\frac{r}{8\pi L_s^2}\right] + \left[g_s^2 + g_s^2\frac{1}{4} + g_s^2\frac{1}{4}\right]\left[\frac{1}{4\pi r}\right]. \tag{9.16}$$

To compare (9.14) and (9.16) with the empirical results for charmonium, we should also include the attractive electromagnetic interaction between a charm quark and antiquark with electric charges $2e/3$ and $-2e/3$, respectively, which contributes $-(2e/3)^2/(4\pi r) = -e^2/(9\pi r)$ to the potential energy. Thus the total potential energy $V_{tp} = V_{t8} + V_{t3} - e^2/(9\pi r)$ for charmonium is given by

$$V_{tp} = g_s^2\left[\frac{r}{8\pi L_s^2}\right] - \frac{3g_s^2}{4\pi r} - \frac{e^2}{9\pi r}. \tag{9.17}$$

The coupling constant g_s and the basic length L_s in the Lagrangian with general Yang-Mills symmetry can be determined using the empirical potential energy V_{ep}. Such a dual quark potential energy V_{ep} was obtained from the analysis of the charmonium data by the Cornell group,[1]

$$V_{ep} = -\frac{\alpha_c}{r}\left[1 - \frac{r^2}{a^2}\right], \quad \alpha_c = 0.2, \quad a = 0.2fm. \tag{9.18}$$

From (9.17) and (9.18), we estimate the total potential energy,

$$V_{tp} \approx \left(\frac{g_s^2}{4\pi}\right)\frac{r}{2L_s^2} - \left(\frac{g_s^2}{4\pi}\right)\frac{3}{r}, \tag{9.19}$$

$$\frac{g_s^2}{4\pi} \approx 0.07, \qquad L_s \approx 0.082 fm,$$

where the electromagnetic coupling strength $e^2 \approx 0.007$ in (9.17) turns out to be negligible. Since the experimental result[1] for m_c is given by $m_c \approx 1.27 \pm 0.02$ GeV rather than $m_c \approx 1.6\ GeV$,[5] the uncertainty of the estimates in (9.19) are roughly 25%, presumably. We have used this estimate of $L_s \approx 0.082 fm$ to find the coupling constant g_b (the baryon charge) for the Okubo force on a cosmic scale in Ch. 3.

It is interesting to note that the symmetry-unified quark-cosmic model for quark confinement with result (9.19) appears to be consistent with perturbation theory because the coupling strength $g_s^2/4\pi \approx 0.07$ is much smaller than 1. We note that within conventional QCD, which does not have an explicit confining potential, the coupling strength is $\alpha_s \approx 0.118$.[1]

9-3.   The simple harmonic oscillator model and proton-neutron masses

Let us now calculate the confining force that this two-quark state (with a spherical density $|\Psi|^2$) exerts on a green d-quark located at a distance $r < R_i$ from the center of two-quark system, where R_i and R_o are, respectively, the inner and outer radii of the spherical shell of hypercharge due to the red and yellow bound quark state. The following calculation resembles the one in Ch. 3, except that the location of the point S (where the green d-quark is located) is inside the spherical shell (see Fig. 9.1).

To calculate the confining force on the green d-quark located at point S inside the shell, we first divide the color hypercharge shell with inner and outer radii of R_i and R_o into infinitesimally thin shells with radii R' and thickness dR', where $R_o > R' > R_i$. We then divide each infinitesimally thin shell into rings, as indicated by the shaded region in Fig. 9.1.

The color hypercharge dC_h of the ring is assumed to be

$$dC_h = \rho_c(2\pi R' sin\theta)(R'd\theta)(dR'), \qquad (9.20)$$

where $2\pi R' sin\theta$ is the circumference of the ring, $R'd\theta$ is the width of the ring, and dR' is the thickness of the ring. The color hypercharge density ρ_c of the red u-quark/yellow u-quark system is

$$\rho_c = [g_s/\sqrt{3}]|\Psi|^2, \qquad \Psi = \frac{1}{a_q^{3/2}/2}e^{-R'/a_q}, \qquad (9.21)$$

where the sum of the color hypercharges of the red and yellow u-quarks is $g_s/(2\sqrt{3}) + g_s/(2\sqrt{3}) = g_s/\sqrt{3}$.[6]

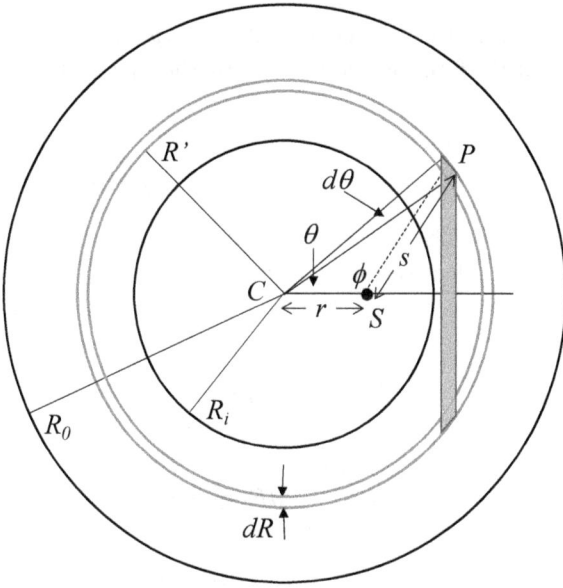

Fig. 9.1. A schematic diagram for calculations of the color hypercharge force between a spherical shell and the green d-quark located at s.

Now let the distance between S and every point of the ring be s and the angle PSC to be ϕ. By symmetry, the confining force dF_{CS} on the green d-quark with the color hypercharge $g_s/\sqrt{3}$ at S by the ring must point along CS.

Note that (9.19) is the attractive potential energy for charmonium $(c\bar{c})$. Replacing the charm quarks by up quarks,[b] we follow the steps from (9.12) to (9.14). In this case, we obtain the total potential energy V'_{tp} and force $F' = -dV'_{tp}/dr$ for the spherical shell between R_i and R_o with color charge $g_s/2\sqrt{3} + g_s/2\sqrt{3}$ and the green d-quark with color charge $-g_s/\sqrt{3}$,

$$V'_{tp} = \frac{g_s^2}{3}\left[\frac{r}{8\pi L_s^2} - \frac{1}{4\pi r}\right], \quad F' = \frac{-g_s^2}{3}\left[\frac{1}{8\pi L_s^2} + \frac{1}{4\pi r^2}\right].$$

This makes the attractive force dF_{CS},

$$dF_{CS} = \frac{-4g_s^2}{3a_q^3}e^{-2R'/a_q}R'^2dR'\int_0^\pi \cos\phi(2\pi \sin\theta d\theta)\left[\frac{1}{8\pi L_s^2} + \frac{1}{4\pi r^2}\right],$$
$$(9.22)$$

---

[b]We do not consider the color isotopic charge because the green d-quark has vanishing color isotopic charge.[6]

where the color hypercharge density $|\Psi|^2$ gives the distribution of color charges in the shell in this model. In (9.22), $cos\phi$ results from taking only the component of the force on the d-quark by the ring that is perpendicular to CS by symmetry.

To evaluate this integral, we note that for triangle CSP, we have

$$s^2 = R'^2 + r^2 - 2R'r \, cos\theta \quad and \quad R'^2 = s^2 + r^2 - 2rs \, cos\phi \qquad (9.23)$$

by the law of cosines. The integration of $d\theta$ from 0 to π (with both R' and r fixed) leads to an effective repulsive force on the d-quark along CS. The magnitude of the total attractive force can be obtained from the integration of ds:

$$dF_{CS} = A \left[e^{-2R'/a_q} \, dR' \, R'^2 \right] \int_0^\pi cos\phi sin\theta \, d\theta \qquad (9.24)$$

$$= A \int_{R'-r}^{R'+r} [dF] \frac{s}{R'r} ds,$$

$$A = \frac{g_s^2}{3L_s^2 a_q^3}, \quad [dF] = e^{-2R'/a_q} \, dR' \, R'^2 \left(\frac{s^2 + r^2 - R'^2}{2sr} \right).$$

Thus, the total attractive confining force $F_{gd}(r) = \int dF_{CS}$ exerted on the green d-quark with hypercharge $-g_s/\sqrt{3}$ is given by

$$F_{gd}(r) \approx A \int_{R_i}^{R_o} e^{-2R'/a_q} \frac{R'dR'}{2r^2} \int_{R'-r}^{R'+r} (s^2 + r^2 - R'^2)ds \qquad (9.25)$$

$$\approx A \int_0^\infty e^{-2R'/a_q} \frac{R'dR'}{2r^2} \left[\frac{8r^3}{3} \right],$$

where we have approximated $R_o \to \infty$ and $R_i \approx 0$ for simplicity. Thus, the confining force acting on the green d-quark $F_{gd}(r)$ is

$$F_{gd}(r) \approx \frac{-g_s^2 r}{9L_s^2 a_q}, \qquad (9.26)$$

where the green d-quark has the color charge $-g_s/\sqrt{3}$. Based on this result, the confining potential $V_{gd}(r)$ of the green d-quark is an SHO potential,

$$V_{gd}(r) = \frac{g_s^2}{18L_s^2 a_q} r^2 = 4b(1.694)(10^8)r^2 \equiv Qr^2 \quad (MeV) \qquad (9.27)$$

$$a_q \equiv L_s/b, \quad 1/fm = 197MeV, \quad L_s = 0.082fm,$$

where b (or equivalently, a_q) is a free parameter to be determined.

As can be seen from the potentials, even before the green d-quark is within the color charge shell of the two u-quark system, there is a long-range attractive force acting on it that increases its probability of entering the shell. Two consecutive 2-body collisions to form a proton have a much larger probability of occurrence than a single 3-body collision.

Based on this model with the potential V_{gd} in (9.27), we can estimate the ground state energy of the 3-body system and compare it with the experimentally measured proton mass. The method of estimation is similar to that in the Bohr atomic model with the classical kinetic energy $p^2/2m_d$ replaced by the relativistic energy $\sqrt{p^2 + m_d^2}$. Using (9.27), the green d-quark has total energy,

$$E_{gd}(p^+) = \sqrt{p^2 + m_d^2} + V_{gd}(r). \tag{9.28}$$

Assuming $\Delta p \approx p$ and $\Delta r \approx r$, we use the uncertainty relation $pr \approx 1$ and $dE_{gd}/dr = 0$ to estimate the minimum or ground state energy of the system. For the relativistic case in (9.28), the condition $dE_{gd}/dr = 0$ gives the following equation for r (in natural units),

$$4Q^2 m_d^2 r^8 + 4Q^2 r^6 - 1 = 0, \quad Q = 4b(1.694 \times 10^8), \tag{9.29}$$

$$4.00535(10^{19})b^2 r^8 + 1.83657(10^{19})b^2 r^6 - 1 = 0,$$

where the numerical values of g_s and L_s are given in (9.19) and the masses of the u-quark and d-quark are $m_u = 2.16 MeV$ and $m_d = 4.67 MeV$, respectively.[2] Solving (9.29) numerically, the only positive real solution is r = 0.001604 for the parameter $b = 0.178$ (determined by trial and error). Using (9.27), (9.28) and (9.21), we estimate the proton mass to be

$$m(p^+) = E_{gd}(p^+) \approx 933.77 MeV + 2m_u = 938.1 \quad MeV, \tag{9.30}$$

where $m_u = 2.16 MeV$. In this 3-quark system, the red and yellow u-quarks are assumed to be in the s-state to form the SHO potential. The kinetic energies of these two u-quarks are relatively small and neglected in (9.30) so that the total energy of the two u-quarks is approximated by $2m_u$.

The neutron mass can be calculated by an analogous procedure. The neutron consists of a red d-quark, a yellow d-quark and a green u-quark. Our model assumes that the two d-quarks form a shell with a SHO potential with the wave function $\propto \Psi$ in (9.22). Following the same steps from (9.22) to (9.27), one obtains the confining potential for the green u-quark, which is the same as (9.27). The green u-quark has the total energy $E_{gu}(n^0)$,

$$E_{gu}(n^0) = \sqrt{p^2 + m_u^2} + V_{gu}(r), \quad V_{gu}(r) = V_{gd}(r), \tag{9.31}$$

where $p \approx 1/r$ and $\Delta p \approx p$ and $\Delta r \approx r$. With the help of the uncertainty relation $pr \approx 1$, the condition $dE_{gu}(r)/dr = 0$ gives

$$4Q^2 m_u^2 r^8 + 4Q^2 r^6 - 1 = 0, \qquad Q = 4b(1.694 \times 10^8), \qquad (9.32)$$

$$2.7149(10^{17})r^8 + 5.81898(10^{16})r^6 - 1 = 0, \qquad b = 0.178.$$

Solving (9.32) numerically, we obtain the neutron mass as

$$m(n^0) = E_{gu}(n^0) \approx 933.758 + 2m_d = 943.1 MeV, \qquad (9.33)$$

where $m_d = 4.67 MeV$, and we used the same value for the parameter $b = L_s/a_q = 0.178$ as for the proton mass in (9.30). The n-p mass difference is approximately

$$m(n^0) - m(p^+) \approx 5.0 MeV. \qquad (9.34)$$

In comparison, the experimental values of the neutron mass and the n-p mass difference are 939.56 MeV and 1.3 MeV,[1] respectively.

The problem of the proton-neutron mass difference has been studied since the 1960s on the basis of the modified $\pi - N$ (pion-nucleon) dispersion relations. The problem could not be solved with traditional methods because a subtraction must be made in the dispersion relations, and the subtraction constant was unknown. One can circumvent this difficulty, but only at the cost of introducing arbitrary parameters. With the modified pion-nucleon dispersion relations, the consistency between the numerical result and the experimental measurement is greatly improved. Nevertheless, the results are very sensitive to the precise choice of parameters in the modified dispersion relations.[8,9] Now that our considerations are based on the experimentally established quarks and confining potential, the results are more reliable from the viewpoint of particle physics.

To summarize, the 3-quark simple harmonic model has the following results:

(a) The total relativistic energy E_{gd} of the green d-quark provides the dominant contribution to the proton mass, as shown in (9.28) and (9.30).

(b) The total relativistic energy E_{gu} of the green u-quark provides the dominant contribution to the neutron mass, as shown in (9.31) and (9.33).

(c) Confions do not contribute to the proton and the neutron masses directly. However, virtual confions are exchanged between quarks to produce the confining force, which leads to the SHO potential (9.27).

(d) According to this simple harmonic model for ground states of 3-quark baryons, a value of 0.178 for the free parameter $b = L_s/a_q$ yields two reasonable results for masses as shown in (9.30) and (9.33).

(e) The 'radius' $a_q = L_s/b \approx 0.46 fm$ in (9.21) could be interpreted as a 'color hypercharge radius' of the proton and the neutron, similar to the experimental proton 'charge radius' of $0.8 fm$.[2]

We hope that these results of the 3-quark simple harmonic model can be tested by high energy experiments in the future.

Of course, for a more precise formulation, we must use the following dynamical equations for confining quarks based on the general Yang-Mills symmetry:

$$i\gamma^{\mu}[\partial_{\mu} + ig_s C_{\mu}^a (\lambda^a/2)]q - Mq = 0, \tag{9.35}$$

$$\partial^2 \partial^{\mu} C_{\mu\nu}^a + (g_s/L_s^2)\bar{q}\gamma_{\nu}(\lambda^a/2)q = 0,$$

$$\frac{g_s^2}{4\pi} \approx 0.07, \qquad L_s \approx 0.082 fm,$$

which are derived from the general $SU_{3c} \times U_{1b}$ invariant Lagrangian (9.1) and the charmonium result in (9.19). One could use numerical methods to solve these equations in the static case with the confining potential (9.27), similar to the analysis of the charmonium spectrum by the Cornell group.[1]

Note added: In Appendix C, (Quantum Shell Model for 3-Quark Bound States), we discuss the dynamical equation (9.35) based on a simplified model with a Dirac Hamiltonian with an effective harmonic oscillator potential to obtain energy eigenvalues for the N baryon spectrum. The highly simplified quantum shell model seems to give a reasonable understanding of all 29 N baryon masses (equations (C25)–(C44)) based on two parameters K and b, together with an ℓ-dependent function. This appears to indicate that there is some value in the simplified quark Hamiltonians H_d in (C17) and H_u in (C3), and in the quantum shell suggested by the wave function $\Psi' \propto e^{-a^2 r^2/2}(a^2 r^2)^{(\ell+k)/2}$, which is a solution of the Sonine-Laguerre equation.

References

1. E. Eichten, K. Gottfried, T. Kinoshita, J. Kogut, K. D. Lane, T.-Y. Yan, Phys. Rev. Lett., **34**, 369 (1975).
2. Particle Data Group, *Particle Physics Booklet*, 2020, p. 166, p. 25.
3. J. P. Hsu, Nuovo. Cim. B **89**, 209, (1985). https://doi.org/10.1007/BF02723545. J. P. Hsu and L. Hsu, *Space-Time, Yang-Mills Gravity and Dynamics of Cosmic Expansion*. (Singapore: World Scientific, 2020). pp. 267–268. See also Refs. 8 and 9.

4. M. Gell-Mann, Physics Letters. 8 (3): 214 (1964).
5. G. Zweig, CERN-TII-401 (1964).
6. K. Huang, *Quarks, Leptons and Gauge Fields*. (Singapore: World Scientific, 2nd ed. 1992). p. 17, p. 256, pp. 252–260.
7. I. M. Gel'fand and G. E. Shilov, *Generalized Functions*, vol. 1, (New York: Academic Press, 1964). p. 280, p. 363.
8. J. P. Hsu, R. N. Mohapatra and S. Okubo, Phys. Rev. **181**, 2011 (1969) and references therein.
9. J. P. Hsu, *Dispersion Sum Rules, Interference Model, Duality, and the p-n Mass Difference*, Ph.D. thesis (U. of Rochester, 1969).

10

A Total Symmetry-Unified Model and Violations of All Internal Gauge Symmetries by Yang-Mills Gravity

10-1. Total symmetry-unified model of all interactions

The electromagnetic and weak interactions have been unified based on a gauge invariant action with the internal gauge groups $SU_2 \times U_1$. This is very satisfactory in the sense that the predictions of quantum electrodynamics (QED), including the $W^\pm$ and Z^o bosons, have been confirmed by experiments. However, this theory could still be improved as it involves two coupling constants, which appear to be intimately related to the gauge bosons, $W^\pm$ and Z^o, instead of one.

The strong interaction is assumed to be described by quantum chromodynamics (QCD) with color SU_3 symmetry. Yet, the conventional formulation does not provide an explicit mechanism for quark confinement. As we have shown in previous chapters however, the general Yang-Mills SU_3 leads to new phase field equations that satisfy fourth-order equations. These give rise to linear and Coulomb-like potentials capable of providing the necessary confinement of quarks. These dual potentials turn out to be just right for understanding the charmonium spectrum, etc., according to the analysis of the Cornell group.[1]

We have shown in previous work[2] that in a total-unified model, there are only approximate internal symmetries, and that the quantum and classical gravitational interaction can be based on a gauge invariant action of an external space-time translation group. In this theory of quantum Yang-Mills gravity based on flat space-time in inertial frames, the space-time coordinates have their usual operational meanings, in contrast to general relativity in which space-time coordinates can only be defined locally in Riemann curved space-time and hence, have no meaning by themselves.[3] Quantum Yang-Mills gravity leads to a Hamilton-Jacobi equation with 'effective metric tensors' in the geometric limit. Such an equation is called an Einstein-Grossmann equation,[4] whose solutions lead to the same experimental predictions as general relativity, including a perihelion shift of the Mercury, the deflection of light by the sun, etc., consistent with experiments carried out in inertial frames.

It appears that all known interactions can be understood on the basis of general Yang-Mills (gYM) symmetries,[5-7] which involve Lorentz vector gauge functions $\omega_\mu^a(x)$ and a Hamilton's characteristic phase factor. The gYM symmetries include the usual gauge symmetries as a special case when ω_μ^a can be expressed as $\partial_\mu \omega^a(x)$ and the Hamilton's characteristic phase factors reduce to the usual phase functions. Therefore, all known interactions can be totally unified based on the gYM symmetry framework, even though they do not have the same coupling constant. Such a unified model could be called a symmetry-unified model of all interactions.

General Yang-Mills (gYM) symmetry assumes a universal principle of dynamic symmetry that any conserved internal charge, which may be only approximately conserved in the presence of Yang-Mills gravity, is associated with a gauge field.

The conserved baryonic charge corresponds to a baryon phase field, which leads to the cosmic Okubo force and could give a field theoretic explanation of the late-time accelerated cosmic expansion, as discussed in Ch. 3. The conserved lepton charge (g_ℓ) is assumed to be associated with the usual U_1 gauge symmetry, which leads to a cosmic Lee-Yang force that, together with the enhanced gravity of high energy neutrinos, can contribute to the observed flat rotation curves of galaxies, as discussed in Ch. 2.

Thus, the forces in nature are gravity, the confining quark forces, the electroweak force, and the cosmic baryonic and cosmic leptonic forces with the following symmetry groups, respectively:

$$T_4 \times SU_{3c} \times (SU_2 \times U_1) \times U_{1b} \times U_{1\ell} = G_{tot}, \qquad (10.1)$$

where G_{tot} denotes the total symmetry group based on the universal principle of dynamic symmetry. The transformations of these groups involve a (Lorentz) vector gauge function, including the T_4 group.

The coupling strengths of these forces are or are estimated to be, respectively

$$\frac{g^2}{4\pi} = 5.36 \times 10^{-76} m^2, \qquad \frac{g_s^2}{4\pi} \approx 0.07, \qquad \left(\frac{f^2}{4\pi} = 0.033, \right. \qquad (10.2)$$

$$\left. \frac{f'^2}{4\pi} = 0.009 \right), \qquad \frac{g_b^2}{4\pi} \approx 10^{-100}, \qquad \frac{g_\ell^2}{4\pi} \approx 10^{-43},$$

where the coupling constants f and f' are defined in the $(SU_2 \times U_1)$ gauge covariant derivative $\partial_\mu + if W_{\mu k} t_k + i f' U_\mu t_o$. (See Eq. (10.7) below.) The values of the coupling constants g_b and g_ℓ in (10.2) are based on extremely rough order-of-magnitude estimates because of the lack of cosmological data, as discussed in Chs. 2 and 3. Even so, these two coupling constants are surely extremely small and their effects cannot be detected by experiments on the Earth. In a sense, it is wonderful that so many different interactions in nature, with such hugely different coupling strengths and playing roles at extremely different scales from the microscopic ($\leq 10^{-15}$ meter) to super-macroscopic ($\geq 10^{25}$ meter) worlds, can be unified under the dynamic general Yang-Mills symmetries (that include the usual gauge symmetries as special cases).

We note three important features of the big picture in such a symmetry-unified model with (10.1):

(A) The gravitational coupling constant g^2 is not dimensionless, in sharp contrast to all other coupling constants, as shown in (10.2). Its value depends on the units, which depend on human conventions. In this sense, its value is probably not an intrinsic property of nature, in contrast to all other dimensionless constants in (10.2), which are independent of human conventions of units. For example, its value in SI units is

$$g^2/4\pi = 5 \times 10^{-37} s/kg.$$

The value of $g^2/4\pi$ in (10.2) is in natural units, which we postulate to be the intrinsic units of nature.[8,9] All physical quantities can then be expressed in terms of a single unit, which could be a length, a time, an energy, etc. Based on natural units, we argue that only dimensionless constants can be intrinsic constants of the physical world.[8] If one chooses a unit of length, such as the meter, as shown in (10.2), it suggests that there is an extremely small fundamental length in the physical world:

$$\sqrt{\frac{g^2}{4\pi}} \approx 2.4 \times 10^{-38} m.$$

This fundamental length could affect quantum physics in the future. In particular, the existence of a fundamental length would imply a minimum uncertainty of position $(\Delta x)_{min}$ and hence, a maximum uncertainty of momentum $(\Delta p)_{max}$. This property suggests (a) the absence of infinite energy and momentum, which could help remove the ultraviolet divergences in quantum field theory,[10] so that all field theories would be finite, as they should be, and (b) that Lorentz-Poincaré invariance could be violated at very large momenta near $(\Delta p)_{max}$.

(B) Although the maximum vertex involves only four gravitons, just like in usual gauge theories, quantum Yang-Mills gravity is not renormalizable by power counting because it has a dimensional coupling constant g. According to the total symmetry-unified model (10.1), since all matter is universally coupled via gravity, all gauge theories that are renormalizable in the absence of gravity will become non-renormalizable once the gravitational interaction is taken into account. In particular, the asymptotic freedom in quantum chromodynamics will no longer be true due to the interaction with quantum Yang-Mills gravity. Strictly speaking, this is not a matter of approximate asymptotic freedom, rather, it is a matter of the lack of asymptotic freedom.

(C) Quantum Yang-Mills gravity provides a new understanding of questions such as why only the gravitational force is always attractive,[4, 10] while all other forces can be both attractive and repulsive, why only the gravitational coupling constant has a dimension, while all other coupling constants are dimensionless, and why there are no forces that are always repulsive. The answers to all of these questions originate from the difference between the generators, $p_\mu = i\partial_\mu$, for the space-time translational group and the generator t_o for the U_1 group. The generators ∂_μ have the representation $i\partial/\partial x^\mu \equiv \partial_\mu$, while the generators of internal groups have constant matrix representations. Since the electromagnetic potential A_μ is a linear combination of $W_{\mu 0}$ and $W_{\mu 3}$, the total gauge covariant derivative acting on a fermion (e.g., electron) field ψ can be written as

$$\mathbf{\Delta}_\mu \psi = (\partial_\mu - ig\phi_\mu^\nu p_\nu + ieA_\mu + ...,)\psi \tag{10.3}$$

$$= (\partial_\mu + g\phi_\mu^\nu \partial_\nu + ieA_\mu + ...,)\psi, \qquad p_\nu = i\partial_\nu.$$

The wave equations of antiparticles involve the complex conjugate of $\mathbf{\Delta}_\mu$. Thus, the gravitational coupling constant for the electron and the positron are the same, while the electromagnetic coupling constants for the electron and the positron differ by a sign.

$$(\mathbf{\Delta}_\mu \psi)^* = (\partial_\mu + g\phi_\mu^\nu \partial_\nu - ieA_\mu + ...,)\psi^*. \tag{10.4}$$

The nature of the electric force between a charged fermion and an antifermion and that between a fermion and another fermion are respectively determined by

$$(ie)(-ie) = e^2 \ (attractive) \quad and \quad (ie)(ie) = -e^2 \ (repulsive). \tag{10.5}$$

On the other hand, the Yang-Mills gravitational force between a fermion and an anti-fermion and that between a fermion and another fermion are respectively given by

$$(+g)(+g) = g^2 \ (attractive) \quad and \quad (+g)(+g) = g^2 \ (attractive). \tag{10.6}$$

Similarly, the gravitational force between an anti-fermion an anti-fermion is also attractive: $(+g)(+g) = g^2$, given by the second term in (10.4).

Since these considerations also apply to other forces associated with internal gauge groups and with dimensionless coupling constants, such as the quark force, baryon force and lepton force, we conclude that within the framework of Broader Particle-Cosmology with general Yang-Mills symmetry (including the usual gauge symmetries as special cases), (10.3)–(10.6) indicate that it is impossible for a purely repulsive force to exist in the physical world.

10-2. Extremely small violations of all internal gauge symmetries by gravity

In the total symmetry-unified model, the total gauge covariant derivative Δ_μ^t in an inertial frame takes the form

$$\Delta_\mu^t \equiv \partial_\mu + g\phi_\mu^\nu \partial_\nu + ig_s C_\mu^a \frac{\lambda^a}{2} + ifW_\mu^k t^k + if'U_\mu t_o \qquad (10.7)$$

$$+ig_b B_\mu + ig_\ell L_\mu,$$

where t_k are SU_2 generators and t_o (weak hypercharge) is the U_1 generator. Such a total unified model suggests that the coupling constants in (10.2) or (10.7) are fundamental and inherent constants of nature, as discussed[8] previously based on natural units.[9]

The gauge curvature associated with each individual group and the Lagrangian for such a total-unified model can be obtained by calculating the commutator of the covariant derivative defined in (10.7). Let us calculate the commutator of the total-unified gauge covariant derivative (10.7),

$$[\Delta_\mu^t, \Delta_\nu^t] = C_{\mu\nu\sigma}\partial^\sigma + ig_s C_{\mu\nu}^a \frac{\lambda^a}{2} + ifW_{\mu\nu}^k t^k + if'U_{\mu\nu} t_o \qquad (10.8)$$

$$+ig_b B_{\mu\nu} + ig_\ell L_{\mu\nu},$$

$$C^{\mu\nu\alpha} = J^{\mu\lambda}(\partial_\lambda J^{\nu\alpha}) - J^{\nu\lambda}(\partial_\lambda J^{\mu\alpha}), \qquad J_\mu^\sigma = \delta_\mu^\sigma + g\phi_\mu^\sigma, \qquad (10.9)$$

$$C_{\mu\nu}^a = \partial_\mu C_\nu^a - \partial_\nu C_\mu^a + g_s f^{abc} C_\mu^b C_\nu^c. \qquad (10.10)$$

The confion has zero mass and zero energy-momentum, similar to a vacuum state (which is not directly observable), so that it is assumed not to couple to gravitons, i.e., $C_{\mu\nu}^a$ is given by (10.10) involving ∂_μ rather than $J_{\mu\lambda}\partial^\lambda$.

The gauge curvatures of $SU_2 \times U_1$, U_{1b} and $U_{1\ell}$ are respectively given by $W^i_{\mu\nu}$, $U_{\mu\nu}$, $B_{\mu\nu}$ and $L_{\mu\nu}$ in the presence of gravity:

$$W^i_{\mu\nu} = J^\sigma_\mu \partial_\sigma W^i_\nu - J^\sigma_\nu \partial_\sigma W^i_\mu - f\epsilon^{ijk} W^j_\mu W^k_\nu, \tag{10.11}$$

$$U_{\mu\nu} = J^\sigma_\mu \partial_\sigma U_\nu - J^\sigma_\nu \partial_\sigma U_\mu, \tag{10.12}$$

$$B_{\mu\nu} = J^\sigma_\mu \partial_\sigma B_\nu - J^\sigma_\nu \partial_\sigma B_\mu, \tag{10.13}$$

$$L_{\mu\nu} = J^\sigma_\mu \partial_\sigma L_\nu - J^\sigma_\nu \partial_\sigma L_\mu, \tag{10.14}$$

where $i, j, k = 1, 2, 3$, for the SU_2 group.[11] The generators of different groups are assumed to be commuting.

In inertial frames with the metric tensor $\eta_{\mu\nu}$, the total symmetry-unified Lagrangian L^{tot} is assumed to be

$$L^{tot} = L_{T_4} + L_{SU_3} + L_{SU_2U_1} + L_{U_{1b}} + L_{U_{1\ell}}, \tag{10.15}$$

$$L_{T_4} = \frac{1}{2g^2} \left(\frac{1}{2} C_{\mu\nu\alpha} C^{\mu\nu\alpha} - C_{\mu\alpha}{}^\alpha C^{\mu\beta}{}_\beta \right), \tag{10.16}$$

$$+ \frac{i}{2} [\overline{\psi} \gamma_\mu (J^{\mu\lambda} \partial_\lambda \psi) - (J^{\mu\lambda} \partial_\lambda \overline{\psi}) \gamma_\mu \psi] - m\overline{\psi}\psi,$$

$$J_{\mu\nu} = \eta_{\mu\nu} + g\phi_{\mu\nu}, \qquad C^{\mu\nu\alpha} = J^{\mu\sigma} \partial_\sigma J^{\nu\alpha} - J^{\nu\sigma} \partial_\sigma J^{\mu\alpha},$$

$$L_{SU_3} = \frac{L_s^2}{2} [\partial^\mu C^a_{\mu\lambda} \partial_\nu C^{a\nu\lambda}] + i\overline{q}[\gamma_\mu J^{\mu\lambda} \partial_\lambda - M]q, \tag{10.17}$$

$$L_{SU_2U_1} = -\frac{1}{4}(W^{\mu\nu a} W^a_{\mu\nu} + U^{\mu\nu} U_{\mu\nu}) + L_{matter}, \tag{10.18}$$

$$L_{U_{1b}} = \frac{L_s^2}{2} (J^\alpha_\lambda \partial_\alpha B_{\mu\nu})(J^{\lambda\beta} \partial_\beta B^{\mu\nu}) \tag{10.19}$$

$$+ i\overline{q}\gamma^\mu (\partial_\mu + g\phi^\nu_\mu \partial_\nu + ig_b B_\mu)q - m_q \overline{q}q,$$

$$L_{U_{1\ell}} = -\frac{1}{4} L_{\mu\nu} L^{\mu\nu} + i\overline{\ell}(x)\gamma^\mu (\partial_\mu + g\phi^\nu_\mu \partial_\nu + ig_\ell L_\mu)\ell(x) - m_\ell \overline{\ell}\ell, \tag{10.20}$$

where $C^{\mu\alpha\beta}$ is the T_4 gauge curvature and the matter Lagrangian L_{matter} in (10.18) includes all fermions and Higgs bosons.

In (10.17), it is understood that the spinor quark fields q with components q^{fi}, where $i = 1, 2, 3$ for color indices and $f = 1, 2, ...6$ for flavor indices (u, d, c, s, t, b) are summed over, and M is a color-independent mass matrix in the flavor indices.[11] The constant L_s denotes a length that characterizes the dynamical systems described by the gYM invariant Lagrangian. We postulate that L_s is the universal length for phase fields with general Yang-Mills symmetry and is an essential feature of the symmetry-unification of all interactions — a common feature of the small quark world and the gigantic cosmic world. Each quark with baryonic charge in (10.19) is treated as a singlet. Similarly, each lepton with leptonic charge in (10.20) is also treated as a singlet. It is understood that all quarks and leptons are summed over. Other terms such as the electromagnetic coupling to charged quarks and leptons in (10.19) and (10.20) are neglected for simplicity, because we are mainly interested in the effect of incorporating Yang-Mills gravity on other interactions.

In the presence of gravity, the U_1 gauge transformations for U_μ and $U_{\mu\nu}$ with the infinitesimal function ω are

$$\delta U_\mu = U'_\mu - U_\mu = \partial_\mu \omega \tag{10.21}$$

$$\delta U_{\mu\nu}(x) = J^\sigma_\mu \partial_\sigma(\delta U_\nu) - J^\sigma_\nu \partial_\sigma(\delta U_\mu) \tag{10.22}$$

$$= g[\phi^\sigma_\mu \partial_\sigma \partial_\nu \omega - \phi^\sigma_\nu \partial_\sigma \partial_\mu \omega] \neq 0.$$

Similarly, the SU_2 gauge transformation of the gauge curvature $W^{\mu\nu}_a$ also differs from that in the absence of gravity,

$$\delta W^{\mu\nu}_a = J^{\mu\sigma}\partial_\sigma(\delta W^\nu_a) - J^{\nu\sigma}\partial_\sigma(\delta W^\mu_a) + f\epsilon_{abc}(\delta W^\mu_b)(\delta W^\nu_c) \tag{10.23}$$

$$\neq [\delta W^{\mu\nu}_a]_{g=0},$$

where δW^μ_a and $[\delta W^{\mu\nu}_a]_{g=0}$ are given by

$$\delta W^\mu_a = \frac{1}{f}\partial^\mu \omega_a(x) + \epsilon_{abc}\omega_b(x)W^\mu_c, \tag{10.24}$$

$$[\delta W^{\mu\nu}_a]_{g=0} = \epsilon_{abc}\omega_b(x)W^{\mu\nu}_c. \tag{10.25}$$

Thus, the Lagrangian $L_{SU_2U_1}$ in (10.18), which involves $W_{a\mu\nu}W^{\mu\nu}_a$ and $U_{\mu\nu}U^{\mu\nu}$, has only approximate $SU_2 \times U_1$ symmetry in the presence of quantum Yang-Mills gravity.

The expressions in (10.15)–(10.25) demonstrate the extremely small violation of all internal gauge symmetries by quantum Yang-Mills gravity and, more importantly, that the space-time translation group is the only 'exact dynamical symmetry group' in the total symmetry-unified model based on Broader Particle-Cosmology.

It is interesting to observe that, for example, the U_1 gauge curvature (or field strength) $L_{\mu\nu}$ in (10.20) involves a very small violation of the U_1 or U_{1b} gauge invariance due to the presence of gravity. In other words, the internal charge of the U_1 symmetry by itself is not exactly conserved. (See Sec. 10-3 below.) Similarly, the U_{1b} gauge curvature $B_{\mu\nu}$ in (10.19) is invariant under U_1 gauge transformations, i.e., $B'_\mu(x) = B_\mu(x) + \partial_\mu\Lambda(x)$, only when $g = 0$ (i.e., in the absence of a gravitational field $\phi_{\mu\nu}$).

10-3. Experimental tests of violation of electromagnetic gauge invariance

The physical effects of the violations of the U_1 gauge symmetry by the Yang-Mills gravity might be detectable in two cases:

(a) an extremely small non-conservation of electric charge, and

(b) a very small non-gauge invariance in electromagnetic phenomena, e.g., the Josephson voltage-phase relation in superconductors.

10-3.1. *Extremely small non-conservation of electric charges*

Let us concentrate on the electromagnetic sector of the symmetry-unified model in (10.15). The action $\int d^4x L_{em}$ involving only the electromagnetic field A_μ and a charged fermion ψ, which are coupled to the gravitational field $\phi_{\mu\nu}$,[1,4] is

$$L_{em} = -\frac{1}{4}F_{\mu\nu}F^{\mu\nu} + \frac{i}{2}\left[\overline{\psi}\gamma^\mu(\Delta_\mu - ieA_\mu)\psi - [(\Delta_\mu + ieA_\mu)\overline{\psi}]\gamma^\mu\psi\right] - m\overline{\psi}\psi,$$
$$(10.26)$$
$$F_{\mu\nu} = \Delta_\mu A_\nu - \Delta_\nu A_\mu, \quad \Delta_\mu = (\delta_\mu^\nu + g\phi_\mu^\nu)\partial_\nu \equiv J_\mu^\nu\partial_\nu, \quad e > 0,$$

in inertial frames with the metric tensors $\eta_{\mu\nu} = (1, -1, -1, -1)$ and $c = \hbar = 1$, where A_μ is assumed to satisfy the usual gauge condition $\partial^\mu A_\mu = 0$. The $W^\pm$ and Z^o gauge bosons and others are not considered here. The unified gauge covariant derivative, $d_\mu = \Delta_\mu - ieA_\mu + ... = \partial_\mu + g\phi_\mu^\nu\partial_\nu - ieA_\mu + ...$, involves the generators of the space-time translational group, $p_\nu = i\partial_\nu$.

The generalized Maxwell's equations in the presence of gravity can be derived from (10.26). We have

$$\partial_\alpha(J_\mu^\alpha F^{\mu\nu}) = -e\overline{\psi}\gamma^\nu\psi, \qquad J_\mu^\alpha = \delta_\mu^\alpha + g\phi_\mu^\alpha. \qquad (10.27)$$

The new Eq. (10.27) implies a modified continuity equation for the electric current in the presence of gravity,

$$\partial_\nu J_{total}^\nu = 0, \qquad J_{tot}^\nu = -e\overline{\psi}\gamma^\nu\psi - g\partial_\alpha(\phi_\mu^\alpha F^{\mu\nu}), \qquad (10.28)$$

where we have used the identity $\partial_\mu\partial_\nu F^{\mu\nu} = 0$.

Thus, the usual current $e\bar{\psi}\gamma^\nu\psi$ in electroweak theory and quantum electrodynamics is no longer exactly conserved in the presence of gravity. Only a total electromagnetic current, composed of both the usual electromagnetic current and a new 'gravity-em current' (corresponding to the second term in (10.28)), is conserved. The total conserved charge in the gravity-electroweak model is given by

$$Q_{tot} = \int J^0_{tot} d^3x = \int [-e\bar{\psi}\gamma^0\psi - g\partial_\alpha(\phi^\alpha_\mu F^{\mu 0})]d^3x. \qquad (10.29)$$

The new effect involves a factor $|g\phi^0_0| \approx |g\phi^1_1| \approx Gm/r$, which is roughly 10^{-9} on the surface of Earth. Such an effect could be tested experimentally if the electric field **E** (i.e., the electromagnetic field strength F_{i0}) is large. For example, one could measure the charge Q_{tot} of a physical system both at the Earth's surface and on the International Space Station, where the gravitational potential due to the Earth is much smaller. The total symmetry-unified model predicts that the measured electric charge Q_{tot} in the two environments will differ by a factor of 10^{-9}. At present, the electron charge magnitude is defined to be $1.602\ 176\ 634 \times 10^{-19}C$.[12] However, making such a definition can only be experimentally consistent if one ignores quantum Yang-Mills gravity or if the electromagnetic U_1 gauge symmetry is an exact symmetry. As indicated in (10.29), quantum Yang-Mills gravity predicts that the electromagnetic gauge symmetry is violated by the presence of gravity, implying that any physical effects related to the electric charge will be different in different gravitational fields, e.g., if one compares an effect measured in an Earth laboratory and the same effect measured on the International Space Station. We hope that the prediction (10.29) by the quantum Yang-Mills gravity can be tested in the near future.

10-3.2. *Experimental tests of the modified Josephson effect*

The gauge-dependent observable results due to an extremely small violation of the electromagnetic gauge invariance suggest a novel experiment with Josephson effects in a superconductor.

Since the exact dynamic equation for a Cooper-pair has not been established, we assume as usual that the quantum equation for Cooper-pairs is more or less like the Schrödinger equation for an electron, with the difference being that the charge Q is assumed to be twice the charge of an electron.[13, 14] The space and time derivatives in the wave equation are modified by the presence of Yang-Mills gravity, as shown in (10.27) and (10.30) below. At the surface of the Earth, the non-vanishing components of $J^\nu_\mu = \eta^{\nu\alpha} J_{\alpha\mu}$ are found to be[4]

$$J^0_0 \approx 1 + g\phi^0_0 \equiv h_0, \quad J^1_1 = J^2_2 = J^3_3 \approx 1 - g\phi^0_0 \equiv h_1; \tag{10.30}$$

$$g\phi^0_0 \approx 6.95 \times 10^{-10},$$

to a first-order approximation. The modified wave equation in the presence of Yang-Mills gravity and the electromagnetic potential (A_0, A_k) is

$$ih_0 \frac{\partial\psi}{\partial t} = \frac{1}{2M}(-ih_1\partial_k - QA_k)^2\psi + QA_0\psi \tag{10.31}$$

where $\partial_k = (\partial_x, \partial_y, \partial_z)$, etc.

In order to estimate the effect of Yang-Mills gravity on the precision DC voltage standard, let us consider the modification of the Josephson voltage-phase relation by gravity, taking into account the tunneling in a basic Josephson junction only. We consider as usual a static case involving the vector potential A_k and a constant voltage V_o in the presence of Yang-Mills gravity,

$$-\frac{1}{2M}(h_1\partial_k - iQA_k)^2\psi(\mathbf{r}) = (h_0E_o - V_o)\psi(\mathbf{r}). \tag{10.32}$$

The wave function $\psi_d(\mathbf{r})$ in the insulating region satisfies[14]

$$-\frac{h^2_1}{2M}\partial^2_k\psi_d(\mathbf{r}) = (h_0E_o - V_o)\psi_d(\mathbf{r}), \quad |x| \le a, \tag{10.33}$$

where $V_o > h_0E_o$, and its solution (for the 1-dimensional case) is

$$\psi_d(x) = G_1\cosh(x/x_g) + G_2\sinh(x/x_g), \tag{10.34}$$

$$x_g = \frac{h_1}{\sqrt{2M(V_o - h_0E_o)}},$$

in the insulating region. As usual, the boundary conditions at $x = -a$ and $x = a$ for the solution in (10.34) are $\psi(-a) = \sqrt{n_2}e^{i\theta_2}$, and $\psi(+a) = \sqrt{n_1}e^{i\theta_1}$, where n_2 (n_1) is the constant Cooper-pair density at $x = -a$ $(+a)$, etc. The current-phase relation modified by Yang-Mills gravity is thus

$$j_g = j_{gc}\sin(\theta_{g21}), \quad j_{gc} = \frac{Qh^2_1\sqrt{n_1n_2}}{Mx_gh_0\sinh(2a/x_g)}, \tag{10.35}$$

$$\theta_{g21} = \theta_2 - \theta_1 - \frac{Q}{h_1}\int_1^2 \mathbf{A}\cdot d\mathbf{s}, \quad \int_1^2 = \int_{+a}^{-a}, \tag{10.36}$$

where j_{gc} is the modified maximum supercurrent density, and the integral of $\mathbf{A}\cdot d\mathbf{s} = A_x dx$ is to be taken across the junction.

Therefore, in the presence of Yang-Mills gravity, the modified Josephson equation for voltage is given by

$$V_{g21} = h_0 \frac{\partial \theta_{g21}}{\partial t} \approx Q \int_1^2 \left[-\nabla A_0 - (h_1)^{-2} \frac{\partial \mathbf{A}}{\partial t} \right] \cdot d\mathbf{s}, \qquad (10.37)$$

which is not gauge invariant. Only in the absence of Yang-Mills gravity do we have $h_0 = h_1 = 1$, in which case the modified voltage V_{g21} in (10.37) reduces to the usual gauge invariant Josephson voltage-phase relation,

$$V_{21} = \frac{\partial \theta_{g21}}{\partial t} = Q \int_1^2 \left[-\nabla A_0 - \frac{\partial \mathbf{A}}{\partial t} \right] \cdot d\mathbf{s} = Q \int_1^2 \mathbf{E} \cdot d\mathbf{s}. \qquad (10.38)$$

The experiment is feasible because one can compare the voltage across a Josephson junction in a laboratory at rest on Earth with that across a junction in free fall (e.g., in the International Space Station). Yang-Mills gravity predicts a difference on the order of 1 part in 10^9, which should be detectable because the precision in the Josephson junction voltage standard[15, 16] is on the order of a few parts in 10^{10}.

Superconducting phenomena involve electromagnetic fields, which are described by scalar and vector potentials.[13] Note that the magnetic flux through a loop is equivalent to the line integral of the vector potential $\mathbf{A}$. If one were to create a rapidly changing magnetic flux, $\partial \mathbf{A}/\partial t$ can be made very large. One can use this change, $\partial \mathbf{A}/\partial t$, to test the last term in (10.37). In general, when the experiment is designed such that the modified voltage (10.37) is dominated by the time derivative of the vector potential $(h_0/h_1)\partial \mathbf{A}/\partial t$, then Yang-Mills gravity predicts that the measured values of V_{g12} on Earth will increase by a factor of $h_0/h_1 \approx (h_1)^{-2} \approx (1+1.4 \times 10^{-9})$ in comparison with those measured in orbit around Earth. On the other hand, if (10.33) is dominated by the first term ∇A_0, the measured values of V_{g12} will not increase.

Such an experiment can actually test two predicted physical effects of quantum Yang-Mills gravity:

(I) The experiment directly tests a specific difference of the voltage V_{g21} on the order of 1 part in 10^9, as shown by $(h_1)^{-2} \approx 1.4 \times 10^{-9}$ in the last term of (10.37) due to gravity.

(II) It also tests indirectly the violations of all internal gauge symmetries — $SU_3, (SU_2 \times U_1), U_{1b}, U_{1\ell}$ by gravity in the symmetry-unified model because they all have the same universal coupling to gravity as shown in (10.7).

If the result of experimental test (I) is positive, it also supports indirectly that the space-time translational gauge symmetry of quantum Yang-Mills gravity is the only exact dynamic symmetry in physics — a most important feature of the symmetry-unified model of all interactions. This exact dynamic symmetry of the space-time translational group is also the underpinning of all discussions in this monograph.

10-4. Summary

It is gratifying that quantum Yang-Mills gravity can be formulated in flat space-time and allows a logically consistent Broad Particle-Cosmology that is based on three underpinnings:

(I) General Yang-Mills symmetry with Yang-Mills gravity and the cosmic isotropy principle lead to a Lagrangian dynamics for the motion of galaxies.

(II) Particle physics gives a symmetry-unified model of all interactions and a CPT-invariant Big Jets model, which predicts the existence of an antimatter half-universe (which in the present is a 3K antimatter-blackbody).

(III) All physics can be formulated in inertial and non-inertial frames with operationally defined coordinates in flat space-time. There are space-time transformations among them, which allow a grid of 'space-time clocks' to be set up in each non-inertial frame. In the limit of zero acceleration, the space-time clocks reduce to the usual synchronized clock system in an inertial frame.

From a cursory glance at the total symmetry-unified model based on the total Lagrangian L^{tot} in (10.15), it may seem that the total covariant derivative (10.7) is simply the juxtaposition of those for various interactions. What is the advantage of such a unification?

One advantage of the symmetry-unification is that one can no longer ignore non-renormalizable quantum Yang-Mills gravity. Because of its universal coupling to all physical fields, two implications becomes clear: All field theories of known interactions in (10.15) are non-renormalizable and hence, the asymptotic freedom of, say, SU_3 gauge theory is no longer true.

We believe that such a total symmetry-unified model may well be the limit of what one can do regarding unification in the physical world as we know it for the following reasons:

(a) Mathematically, it appears very unlikely that one can further unify all these external and internal gauge symmetries into one single Lie group with one single coupling strength.

(b) Physically, the usual asymptotic freedom and running coupling constants in non-abelian gauge theories are only approximately true because of Yang-Mills gravity.

In the symmetry-unified framework including gravity, any quantum field theory needs an infinite numbers of counter terms to define the theory at all orders. The conventional view is that a non-renormalizable theory cannot be defined by a finite number of counter terms in the action and hence, is unacceptable.

However, the approach of symmetry-unification suggests a new view that physicists could consider regarding the 'acceptability' of a field theory: Suppose a field theory has an action consisting solely of terms allowed by general Yang-Mills symmetry. And suppose the divergent terms in the finite-loop diagrams can be completely removed by counter terms in the action. Then for all practical purposes, such a theory with finite-loop renormalization is essentially equivalent to a renormalized gauge theory.[17, 18]

The underlying reasons for the ultraviolet divergence in field theory are roughly as follows: A physical particle is probably not a geometric point. However, we do not yet have the rigorous mathematics necessary to deal with non-local particles and fields, as one can see from the heroic efforts of Yukawa and his collaborators.[19] Thus we are forced to deal with local fields with ultraviolet divergences. It is fair to say that the renormalization procedure to make a divergent field theory finite is, strictly speaking, not logical within the framework of local field theories.[20] However, renormalization appears to be an ingenious feat of physicists to extract experimentally consistent numbers from a messy sea of infinities.

Finally, symmetry-unification brings out a key property of interactions. Namely, it appears extremely unlikely that the hugely different interaction strengths from 10^{-3} to $\approx 10^{-100}$ as estimated in (10.2) could be reasonably unified at high energies based on one single coupling constant and one Lie group. Therefore, the conventional idea of unifications of all interactions remains only a vision. The new idea of symmetry-unification of all interactions sheds light on the key idea of this monograph:

General Yang-Mills symmetry appears to be an effective key to unlock the mysteries of interactions and is the heartbeat of the universe.

References

1. E. Eichten, K. Gottfried, T. Kinoshita, J. Kogut, K. D. Lane, T.-Y. Yan, Phys. Rev. Lett., **34**, 369 (1975).

2. J. P. Hsu, Chinese J. Phys. **52**, 692 (2014); Int. J. Mod. Phys. A, **21**, 5119 (2006); J. P. IIsu, Int. J. Mod. Phys. A, **24**, 5217 (2009).

3. E. P. Wigner, *Symmetries and Reflections, Scientific Essays*, (The MIT Press, 1967), pp. 52–53 (in collaboration with H. Salecker). Wigner wrote: 'Evidently, the usual statements about future positions of particles, as specified by their coordinates, are not meaningful statements in general relativity. This is a point which cannot be emphasized strongly enough ....' S. S. Chern (with W. H. Chen), *Lecture Notes on Differential Geometry*, (Lian Jing Publisher, Taipei, 1987) p. iv.

4. J. P. Hsu and L. Hsu, *Space-Time, Yang-Mills Gravity and Dynamics of Cosmic Expansion.* (Singapore: World Scientific, 2020). pp. 122–124, p. 133.

5. J. P. Hsu, Mod. Phys. Lett. A **29**, 1450031 (2014).

6. J. P. Hsu, Eur. Phys. J. Plus, **129**, 108 (2014).

7. J. P. Hsu, Mod. Phys. Lett. A **31** 1650200 (2016).

8. L. Hsu and J. P. Hsu, Eur. Phys. J. Plus **127**, 11 (2012), DOI 10.1140/epjp/i2012-12011-5; J. P. Hsu, Chin. Phys. C, 41, 015101 (2017).

9. W. Pauli, Phys. Rev. **58**, 716 (1940).

10. J. P. Hsu, Chinese Phys. C **43**, 105103 (2019). arXiv: 1908.01585v1; Nuovo Cimento **78 B**, 85 (1983).

11. K. Huang, *Quarks, Leptons and Gauge Fields.* (Singapore: World Scientific, 2nd ed. 1992). pp. 17–33, 252–260.

12. Particle Data Group, *Particle Physics Booklet*, 2020, p. 6.

13. R. Feynman, 'A Seminar on Superconductivity' in The Feynman Lecture on Physics, vol. III, Ch. 21, 6/22–7/22.

14. JOSEPHSON JUNCTION, Massachusetts Institute of Technology. 6.763 2003 Lecture 11. (online, accessed 3 Jan. 2022).

15. M. Grundmann, 'Dimensionality' in Encyclopedia of Condensed Matter Physics, 2005. Josephson Junctions. (accessed 17 Jan. 2022) 'A Josephson junction standard can realize a voltage with an accuracy of 10^{-10}.'

16. D. G. McDonald, Physics Today, July 2001, p. 46.

17. For finite-loop renormalization, see pp. 191–192 in Ref. 4.

18. Similar ideas have been used and discussed by physicists. See, for example, B. L. Voronov and I. V. Tyutin, There. Math. Phys. **50**, 218 (982); **52**, 628 (1982); M. Harada, T. Kugo and Y. Yamawaki, Prog. Theor. Phys. **91**, 801 (1994); J. Gomis and S. Weinberg, in *The Quantum Theory of Fields* (Cambridge Univ. Press, 1995) Vol. 1, pp. 515–525, Vol. 2, pp. 91–95.

19. Y. Katayama, in Proc. 1967 Int. Conf. Paticles and Fields (Eds. C. R. Hagen, G. Guralnik and V. A. Mathur, Interscience, 1967) p. 157.

20. P. A. M. Dirac, Lectures on Quantum Field Theory (Academic Press, 1967) p. 2.

Appendix A

Quantum Yang-Mills Gravity vs. Classical Einstein Gravity

A-1. Great leaps forward at the dawn of the 20th century

In our understanding of the physical world in the past millennium, space-time has emerged as a fundamental framework only in more recent times. It all started in 1905 when two momentous works were completed by the great mathematician Poincaré, 'On the Dynamics of the Electron,' and by the ingenious young physicist Einstein, 'On the Electrodynamics of Moving Bodies.'[1] Both of these works are based on the principle of relativity and a universal speed of light[a] in inertial frames, whose coordinates are related by the Lorentz transformations in flat space-time. Poincaré's work was logical and rigorous, including the derivation of general symmetry group properties of the Lorentz transformations for inertial frames with relative velocities in arbitrary directions. As a mature mathematician, when he discussed 'The Lorentz Group' he used a simple phrase to characterize the Lorentz transformations:[b]

"a linear transformation that leaves invariant the quadratic form $\mathbf{x}^2 - t^2$."

His derivation was short but mathematically flawless.[c] Many physicists at the time did not recognize that Poincaré had derived the space-time transformations.[d] Being a firm believer in the aether, Poincaré invented a strange mechanism to explain why motion relative to the aether cannot be detected in order to have relativity.

The young Einstein's work described all aspects of relativity, except that he obtained only the approximate equation of motion for an electron

---

[a]Historically, a universal speed of light was not proposed until 1887 by Voigt[1] in his discussion of the Doppler effect based on the invariance of the law of the propagation of light in aether!

[b]The same exact transformations were actually first given by Larmor in 1899 in his book 'Aether and Matter.'[1]

[c]Poincaré defined the speed of light to be 1, similar to the natural units $c = \hbar = 1$ used by theorists nowadays.

[d]Nowadays, many textbook authors use Poincaré reasoning to derive the Lorentz transformations, with the explanation that, for example, $\mathbf{x}^2 - c^2 t^2 = 0$ is the law describing the propagation of light and must be invariant, as postulated by the principle of relativity.

with small accelerations and hence, its equation of motion is not invariant under space-time transformations. In 1906, Einstein's work was improved by Planck, who noted the defect and used the invariant principle of least action[e] to derive the exact Lorentz invariant equation for the motion of an electron with arbitrary accelerations. Einstein used about five pages to explain and derive the space-time transformations by considering the emission and propagation of light with a universal speed. In his own way, he clarified the physical meaning of the relativity of time and the universality of the speed of light. He also discussed an experimental test of his relativity theory using Doppler effects. The young Einstein was not hampered by the old baggage of the aether. He simply believed in relativity as a general and fundamental principle not in need of explanation. What a refreshingly innocent and profound viewpoint! Lorentz and others were greatly impressed. Einstein's approach and results captured physicists' imagination and made a great impact.

In 1905, Einstein considered the aether to be superfluous. Nevertheless, many physicists discussed the physical properties of the vacuum related to gauge field theories in the 1970's. They simply assumed that the vacuum is Lorentz-invariant, even though vacuum is very complicated from the viewpoint of modern quantum field theory.[1]

Three years later the mathematician Minkowski expounded the profound idea of the basic four-dimensional space-time[f] symmetry framework, which paved the way mathematically for Einstein's and Grossmann's work on the general theory of relativity[g] in 1913, for Hilbert's Lagrangian to derive the Einstein equation in 1915[2] and for relativistic field theories in the decades to come.

It appears that Pauli's 'Theory of Relativity' published in 1921 had a great influence on physicists' and mathematicians' view of Einstein gravity: "This fusion of two previously quite disconnected subjects — metric and gravitation — must be considered as the most beautiful achievement of the general theory of relativity."[3] Such a profound idea touched the heart of most physicists and mathematicians. For the first time, it provided a new perspective and framework that eventually allowed physicists to explore cosmic phenomena and to model the universe.

However, as early as 1917, the character of the postulate of invariance

---

[e]This invariant principle was used by Poincaré in his 1905 work.
[f]This idea was noted by Poincaré in his 1905 work.
[g]It was realized later that actually, there is no relativity in the general theory of relativity because the framework of General Relativity does not allow for inertial frames and Lorentz transformations in flat space-time.

with respect to general coordinate transformations in general relativity as a geometrical invariance was questioned already by E. Kretschman.[3] Similarly, in 1918 Pauli wrote in his well-known book *Theory of Relativity*,[3] "Lenard raises doubts on the use of coordinate systems of such generality and on the reality of the gravitational fields which would appear in them according to Einstein's theory. The author cannot agree with these objections." Over the next 100 years, most physicists appear to have followed Pauli's viewpoint of Einstein gravity and to have ignored Lenard's penetrating observations and doubts. However, Lenard's objections have not been resolved.

The law of energy conservation in Einstein gravity was investigated by Noether from the viewpoint of group theory in mathematics. In 1918, she proved the well-known theorem about the conservation laws and the invariance of a Lagrangian (called Theorem I). In the same paper, she also proved a Theorem II, which implied that energy could be conserved if and only if a physical theory is invariant under a finite or countably infinite number of infinitesimal generators. However, Einstein gravity does not have a conservation law of energy because the theory is based on a Lie group formed by general coordinate transformations with a continuously infinite number of generators.[4] In this connection, we note that the conserved energy-momentum in Einstein gravity turns out to be a pseudo-tensor,[5] which does not have a physical meaning.[h]

Wigner wrote in the 1960s: "Evidently, the usual statements about future positions of particles, as specified by their coordinates, are not meaningful statements in general relativity. This is a point which cannot be emphasized strongly enough and is the basis of a much deeper dilemma than the more technical question of the Lorentz invariance of the quantum field equations. It pervades all the general theory, and to some degree we mislead both our students and ourselves when we calculate, for instance, the mercury [sic] perihelion motion without explaining how our coordinate system is fixed in space........ Expressing our results in terms of the values of coordinates became a habit with us to such a degree that we adhere to this habit also in general relativity, where values of coordinates are not per se meaningful."[7,i]

---

[h]Because the energy-momentum tensor in Einstein's equation satisfies a covariant continuity equation, some physicists use (or abuse) it as a conservation law in general. However, the covariant continuity equation implies the conservation of energy-momentum only in the very special case of space-time with constant curvature.[6]

[i]It appears that many physicists could not see or believe Kretschman's question, Lenard's doubts or Wigner's critical comments.

It has been taken for granted that the space-time coordinates of any physics theory must have operational meaning, so that the coordinates of any event in a reference frame can be measured and the predictions of the theory can be tested experimentally. This has been clear from the beginning of the twentieth century when special relativity was created. So, it is intriguing that the general theory of relativity with arbitrary coordinates was also created by Einstein, who created special relativity based on an emphasis of the operational meaning of space and time coordinates.

Mathematically, the Riemann geometry of curved space-time is the base of general relativity. However, coordinates in Riemann geometry or a manifold[j] are local and have no meaning per se.[8] This is an inherent property of local coordinates in a Riemann manifold. It appears that there is no possible solution to Wigner's comments and doubts as long as one works within the framework of curved space-time or the Riemann manifold. This has been a long standing problem.

Noether's theorem II mentioned earlier appears to be related to Wigner's insight about transformations and invariance discussed in his Nobel Lecture in 1963: One must have "active transformations, replacing events A, B, C,... by events A', B', C',... and unless active transformations are possible, there is no physically meaningful invariance. However, the mere replacement of one curvilinear coordinate system by another is a 'redescription' in the sense of Melvin;[7] it does not change the events and does not represent a structure in the laws of nature."

It appears that all these properties and problems of general relativity, or Einstein gravity, are related to its mathematical framework of curved space-time with local coordinates.

In a J. W. Gibbs Lecture, given under the auspices of the American Mathematical Society (1972), Dyson said: 'There is no part of physics more coherent mathematically and more satisfying aesthetically than this classical theory of Einstein based upon E-invariance.'[k] However, he also observed:[9] "The most glaring incompatibility of concepts in contemporary physics is that between Einstein's principle of general coordinate invariance and all the modern schemes for a quantum-mechanical description of nature."

---

[j]One can introduce a system of local coordinates in the neighborhood of every point in a manifold. Roughly speaking, the manifold consists of many pieces of Euclidean space patched up together.

[k]That is, based upon general coordinate invariance.

Thus, after the creation of Einstein's general theory of relativity in 1916, the fundamental framework for physics splits into two parts: one for gravity and one for everything else. One way to resolve the problem and to provide a unified fundamental framework for all physics is to formulate a theory of gravity based on the Yang-Mills gauge symmetry in flat space-time. In the 2000's, we proposed and discussed a new classical and quantum Yang-Mills gravity based on local translational gauge symmetry in flat space-time.[10] Yang-Mills gravity is consistent with experiments, and it gives an elegant explanation as to why the gravitational force is always attractive, in contrast to the electromagnetic force. (See Eqs. (1.11)–(1.16) in Ch. 1.) It provides solutions to long-standing difficulties in physics, such as 'the most glaring incompatibility' stressed by Dyson and the failure of energy conservation implied by Noether's theorem II. The problems of quantization of the gravitational field, the operational meaning of space-time coordinates and momenta and the conservation of energy-momentum are all resolved in Yang-Mills gravity. The mathematical basis of Yang-Mills gravity turns out to be the theory of Lie derivatives in coordinate expressions within flat space-time, rather than in the Riemann geometry of curved space-time.[10]

A-2. Similarities in mathematical expressions of Yang-Mills gravity in flat space-time and Einstein gravity in curved space-time

Foundational physics is defined as the invariance of the action under local gauge transformations. Different choices of gauge groups lead to different field theories. If one chooses the internal gauge group SU_3, then one has quantum chromodynamics. If one chooses the external space-time translation group, then one has quantum Yang-Mills gravity;[10] and if one chooses the general coordinate transformation group, then one has classical Einstein gravity.

It turns out that these two different formulations of gravity are based formally on the same mathematical expression for space-time transformations:

$$x^\mu \to x'^\mu = x^\mu + \Lambda^\mu(x), \quad x \equiv x^\lambda, \tag{A1}$$

where $\Lambda^\mu(x)$ is an arbitrary infinitesimal vector function (for simplicity) that can be identified mathematically with the vector field in the Lie derivative $\mathcal{L}_\Lambda$ of arbitrary tensors.

Let us briefly consider two different viewpoints of the space-time transformations (A1):

(A) *Yang-Mills gravity.*   In Yang-Mills gravity, $\Lambda^\mu(x)$ in (A1) are interpreted as local translations in flat space-time. The Yang-Mills gauge covariant derivative Δ_μ involves representations of the generators $p^\mu = i\partial^\mu$, $(c = \hbar = 1)$, of the space-time translation (T_4) group,

$$\Delta_\mu\psi(x) = \partial_\mu\psi(x) - ig\phi_{\mu\nu}(x)p^\nu\psi(x) = [\partial_\mu + g\phi_{\mu\nu}(x)\partial^\nu]\psi(x), \qquad (A2)$$

which dictates the universal coupling between the gravitational tensor field $\phi^{\mu\nu} = \phi^{\nu\mu}$ and the fermion fields ψ of, say, leptons or quarks. At the same time, equations (A1), with a suitable choice of the function $\Lambda^\mu(x)$, can be considered as the coordinate transformations between inertial frames with the Minkowsky metric tensor $\eta_{\mu\nu} = (1, -1, -1, -1)$.[1] In Yang-Mills gravity, x^μ are coordinates in inertial frames and have well-defined operational meanings, as required by quantum field theories with Poincaré invariance. When one considers the Lie derivative in coordinate expressions, they can be understood in general as the coordinates in inertia and non-inertial frames[m] in flat space-time.

(B) *Einstein gravity.*   In General Relativity, (A1) is interpreted as general coordinate transformations in curved space-time, which include all one-to-one and twice-differentiable transformations of the coordinates.[9] This interpretation follows from the principle of general coordinate invariance, i.e., the laws of physics should be invariant under the general coordinate transformations (A1).[6,7] This principle implies that space-time coordinates can be given arbitrary values. For example, clocks in the clock system of general relativity can have arbitrary rates of ticking.[5] In curved space-time, one has the Riemannian covariant derivative, e.g.,

$$D_\mu V^\nu(x) = \partial_\mu V^\nu(x) + \Gamma^\nu_{\mu\lambda}(x)V^\lambda(x), \qquad (A3)$$

where $\Gamma^\nu_{\mu\lambda}(x)$ is the torsionless Levi-Civita connection. The vector field $V^\nu(x)$ in expression (A3) can be generalized to arbitrary tensor fields. The coordinates x^μ in (A3) are local in curved space-time or a Riemannian manifold and have no meaning per se.[8]

We note that both covariant derivatives (A2) and (A3) can be considered as two special cases of covariant derivatives in the theory of Lie derivatives in coordinate expressions.[11,12] The T_4 covariant derivative (A2) is new in the sense that it is related to the 'new interpretation' of the transformations (A1) as the local flat space-time translation T_4 group.[11] In contrast, the

---

[1]Or in non-inertial frames with the Poincaré metric tensor $P_{\mu\nu}$, which reduces to the Minkowsky metric tensor in the limit of zero acceleration.[10]
[m]In this case, the partial derivative ∂_μ in (A2) is replaced by the covariant partial derivative involving the Poincaré metric tensors in flat space-time.

usual covariant derivative (A3) in the theory of Lie derivatives involves any symmetric (or torsion-free) covariant derivative.[12] In other words, the covariant derivative (A3) in Einstein gravity can be a special case in the theory of Lie derivatives that involves the Levi-Civita connection in curved space-time.

We stress that the T_4 space-time translational gauge transformations of arbitrary tensors with infinitesimal vector gauge function $\Lambda^\mu(x)$ in Yang-Mills gravity is exactly the same as the Lie derivatives $\mathcal{L}_\Lambda$ of arbitrary tensors in the coordinate expressions in flat space-time.[11] The arbitrary gauge function $\Lambda^\mu(x)$ in (A1) can be identified with the vector function in the Lie derivatives $\mathcal{L}_\Lambda$. Furthermore, the T_4 gauge invariance of an action in Yang-Mills gravity is the same as the vanishing of the Lie derivative of the action. H. Cartan's formula[13,n] facilitates the calculation of the change of the volume $W d^4 x$ and the invariance of the action in Yang-Mills gravity under the T_4 gauge transformations.[11] Thus, the theory of Lie derivatives (in coordinate expressions) could be considered to be the mathematical basis of Yang-Mills gravity.[o]

It is interesting that mathematically, the Lie derivative has broad flexibility. It can be defined without having to specify a coordinate system, or it can also be defined with a coordinate system in flat space-time with inertial (and non-inertial) frames. The covariant derivative (A3) in Einstein gravity is coordinate-independent, while (A2) in Yang-Mills gravity is coordinate-dependent. This appears to suggest that Einstein gravity being coordinate-independent and Yang-Mills gravity being coordinate-dependent are formally closely related. In other words, 'IF' one treats x^μ at the end of calculations in Einstein gravity as the coordinate x^μ of an inertial frame, the numerical results turn out to be similar to that of Yang-Mills gravity in inertia frames. Clearly, there is no logical connection between Yang-Mills gravity and Einstein gravity. Their formal similarity is probably related to the transformations (A1) and the mathematical flexibility and generality of the Lie derivative.

Based on Yang-Mills gravity, the equation of motion of a classical object (or a light ray) can be derived from the geometric-optics limit of the Dirac equation (or Maxwell equations).[10] We obtain

$$G^{\mu\nu}(x)\partial_\mu S\, \partial_\nu S - m^2 = 0, \qquad G^{\mu\nu}(x) = \eta_{\alpha\beta} J^{\alpha\mu}(x) J^{\beta\nu}(x), \qquad \text{(A4)}$$

---

[n]H. Cartan is the son of E. Cartan.

[o]Historically, Pauli appears to have been the first to discuss a new variation $\delta^* a^\mu = a'^\mu(x) - a^\mu(x)$ for all tensors in his book on relativity (p. 66) published in 1921.[3] In 1931, Ślebodziński[14] introduced a new differential operator for all tensors in his discussion of Hamilton's equations. Later, it was named the Lie derivative.[11]

where S is defined through the limiting expression for, say, the fermion field $\psi = \psi_o exp(iS)$. We call this macroscopic equation (A4), which is derived in Yang-Mills gravity and involves an effective metric tensor $G^{\mu\nu}$, the Einstein-Grossmann equation of motion for classical objects in flat space-time to recognize their collaboration.

The non-linear T_4 gravitational field equation can be linearized, which can then be simplified in the form:

$$\partial_\lambda \partial^\lambda \phi^{\mu\nu} - \partial^\mu \partial_\lambda \phi^{\lambda\nu} + \partial^\mu \partial^\nu \phi^\lambda_\lambda - \partial^\nu \partial_\lambda \phi^{\lambda\mu} = g\left(S^{\mu\nu} - \frac{1}{2}\eta^{\mu\nu}S^\lambda_\lambda\right), \quad (A5)$$

where we have set the gauge parameter $\xi = 0$ and used $J^{\mu\nu} = \eta^{\mu\nu} + g\phi^{\mu\nu}$. It is interesting that the linearized gauge-field Eq. (A5) is mathematically the same as the corresponding equation in Einstein gravity,[P] which may be related to the fact that the space-time translation gauge transformation (A1) in Yang-Mills gravity is formally the same as that in Einstein gravity.

To see the relationship between (A5) and the effective metric tensor $G^{\mu\nu} = G^{\mu\nu}(x)$

$$G^{\mu\nu} = \eta_{\alpha\beta}(\eta^{\alpha\mu} + g\phi^{\alpha\mu})(\eta^{\beta\nu} + g\phi^{\beta\nu}). \quad (A6)$$

To the first order in g, we would like to demonstrate that the effective metric tensor $G^{\mu\nu}$ in (A6) obeys the same equation in YM gravity with the field equation:

$$\partial_\lambda \partial^\lambda G^{\mu\nu} = 2g\partial_\lambda \partial^\lambda \phi^{\mu\nu}, \qquad \partial^\mu \partial_\lambda G^{\lambda\nu} = 2g\partial^\mu \partial_\lambda \phi^{\lambda\nu},$$

$$\partial^\mu \partial^\nu G^\lambda_\lambda = 2g\partial^\mu \partial^\nu \phi^\lambda_\lambda, \qquad \partial^\nu \partial_\lambda G^{\lambda\mu} = 2g\partial^\nu \partial_\lambda \phi^{\lambda\mu}. \quad (A7)$$

From (A5)–(A7), we have the equation for the effective metric tensor $G^{\mu\nu}$ in Yang-Mills gravity:

$$\partial_\lambda \partial^\lambda G^{\mu\nu} - \partial^\mu \partial_\lambda G^{\lambda\nu} + \partial^\mu \partial^\nu G^\lambda_\lambda - \partial^\nu \partial_\lambda G^{\lambda\mu} = 2g^2\left(S^{\mu\nu} - \frac{1}{2}\eta^{\mu\nu}S^\lambda_\lambda\right) + O(g^3),$$
$$(A8)$$

where $2g^2 = 16\pi G$. The factor $2g^2$ on the right side of (A8) is related to the equation $G^{\mu\nu} = \eta^{\mu\nu} + 2g\phi^{\mu\nu} + O(g^2)$.

[P]See Eq. (11.94) on p. 324, $R_{ik} = (1/2)(-g^{(0)lm}\partial_l\partial_m h_{ik} + \partial_k\partial_l h^l_i + \partial_i\partial_l h^l_k - \partial_i\partial_k h^l_l)$, $h_{ik} = g_{ik} - g^{(0)}_{ik}$, and Eq. (11.36) on p. 299, $R_{ik} = (8\pi k/c^4)(T_{ik} - (1/2)g_{ik}T)$, in Ref. 5.

Equation (A8) for the effective metric tensor in Yang-Mills gravity turns out to be the same as the Einstein equation for the metric tensor $g_{\mu\nu}$ to the lowest order in G:[5]

$$\partial_\lambda \partial^\lambda g^{\mu\nu} - \partial^\mu \partial_\lambda g^{\lambda\nu} + \partial^\mu \partial_\nu g^\lambda_\lambda - \partial^\nu \partial_\lambda g^{\lambda\mu} = 16\pi G \left(T^{\mu\nu} - \frac{1}{2}\eta^{\mu\nu}T^\lambda_\lambda \right), \quad (A9)$$

where $T^{\mu\nu}$ is the energy-momentum tensor in Einstein gravity. Thus, the classical field Eqs. (A8) and (A9) imply that in the weak gravity limit, Yang-Mills gravity and Einstein gravity are identical, provided their space-time coordinates x^μ are identified with those in an inertial frame and $S^{\mu\nu} = T^{\mu\nu}$ for macroscopic objects. This property is, of course, consistent with the view that Einstein gravity appears to be an 'effective gauge field theory,' as we will discuss below.

In the following Table A.1, we display the formal similarities in the mathematical expressions of Yang-Mills gravity and Einstein gravity.

Table A.1

Yang-Mills Gravity	Einstein Gravity
$x^\mu \to x'^\mu = x^\mu + \Lambda^\mu(x), \quad x = x^\lambda,$	$x^\mu \to x'^\mu = x^\mu + \Lambda^\mu(x),$
flat space − time,	*curved space − time,*
T_4 *tensor field* : $\phi_{\mu\nu}(x),$	*metric field* : $g_{\mu\nu}(x),$
coordinates have operational meaning,	*local coordinates have no meaning,*
inertial frame,	*reference frame undefined*,*
T_4 gauge covariant derivative: $[\partial_\mu + g\phi_{\mu\nu}(x)\partial^\nu]\psi(x),$	covariant derivative: $\partial_\mu V^\nu(x) + \Gamma^\nu_{\mu\lambda}(x)V^\lambda(x),$
linearized equation: $[\partial_\lambda \partial^\lambda \phi^{\mu\nu} - \partial^\mu \partial_\lambda \phi^{\lambda\nu} + \partial^\mu \partial^\nu \phi^\lambda_\lambda - \partial^\nu \partial_\lambda \phi^{\lambda\mu}] \approx g\left[S^{\mu\nu} - \frac{1}{2}\eta^{\mu\nu}S^\lambda_\lambda\right],$	linearized equation: $[\partial_\lambda \partial^\lambda g^{\mu\nu} - \partial^\mu \partial_\lambda g^{\lambda\nu} + \partial^\mu \partial^\nu g^\lambda_\lambda - \partial^\nu \partial_\lambda g^{\lambda\mu}] \approx g\left[T^{\mu\nu} - \frac{1}{2}\eta^{\mu\nu}T^\lambda_\lambda\right],$
Hamilton-Jacobi type eq.: $G^{\mu\nu}(\partial_\mu S)(\partial_\nu S) - m^2 = 0,$	Hamilton-Jacobi type eq.: $g^{\mu\nu}(\partial_\mu S')(\partial_\nu S') - m^2 = 0,$**
effective metric tensor: $G^{\mu\nu} \equiv \eta_{\alpha\beta}(\eta^{\alpha\mu} + g\phi^{\alpha\mu})(\eta^{\beta\nu} + g\phi^{\beta\nu}),$**	Riemann metric tensor $g^{\mu\nu}$

YM equation for perihelion shift:	Einstein eq. for perihelion shift:
$d^2\sigma/d\phi^2 = \frac{1}{P} - \sigma(1+Q)$***$+ 3Gm\sigma^2$,	$d^2\sigma/d\phi^2 = \frac{1}{P} - \sigma + 3Gm\sigma^2$,
deflection of light (by the sun, $\approx 1.75''$),	*deflection of light ($\approx 1.75''$)*,
red shifts ($\omega_2/\omega_1 \approx 1 + g\phi_1^{00} - g\phi_2^{00}$),	*red shifts (same)*,
*quadrupole radiation $(32G\Omega^6 I^2 e_q^2/5)$****,	*quadrupole radiation (same)*,

---

*Space-time coordinates x^μ are 'effectively' identified with those in an inertial frame in all applications and experimental tests.

**Formally the same linearized field equations and Hamilton-Jacob equations for classical objects in Yang-Mills gravity and Einstein gravity suggests that all gravitational effects are the same in both theories of gravity to a first order approximation, provided the coordinate x_μ in Einstein gravity is identified with that in an inertial frame. These properties have been substantiated by explicit calculations.

***In the equation for the Mercury perihelion shift, $\sigma = 1/r$ and m is the solar mass. There is an additional term Q in Yang-Mills gravity, where $Q = 6Gm(E_o^2 - m_p^2)/(Pm_p^2)$, $P = M^2/(Gmm_p^2)$ and M is the angular momentum of the planet Mercury (with mass m_p). We estimate that the second order difference Q (involving G^2) $\approx 10^{-12}$, which is negligible.

****To a second order approximation for this case, it turns out to give the same total power $P_o(\omega)$ emitted by a body rotating around one of the principal axes of the ellipsoid of inertia (at twice the rotating frequency, $\omega = 2\Omega$). The moment of inertia and equatorial ellipticity are I and e_q, respectively.[10]

A-3.   Einstein Gravity as an 'Effective Gauge Field Theory'

Einstein gravity is based on the principle of general coordinate invariance. This principle implies that the coordinates x^μ in the gravitational field equation can be given any arbitrary values.[7] Thus, reference frames are not defined in Einstein gravity, and the space and time coordinates have no operational meaning. However, it is intriguing that general relativity (GR) gives an equation of motion for a classical object. For example, its solution for the Mercury perihelion shift of 1.75" per century turns out to be consistent with experiments carried out in inertial frames, provided the coordinates x^μ in GR are identified with those in an inertial frame.

In a lucid discussion of the lack of operational meaning of coordinates and momenta in Einstein gravity, Wigner wrote[7] 'The basic premise of this theory [the general theory of relativity] is that coordinates are only auxiliary quantities which can be given arbitrary values for every event. Hence, the measurement of position, that is, of the space coordinates, is certainly not a significant measurement if the postulates of the general theory are adopted: the coordinates can be given any value one wants. The same holds for momenta. Most of us have struggled with the problem of how, under these premises, the general theory of relativity can make meaningful statements and predictions at all.'

In view of the intriguing consistency between general relativity (GR) and experiments carried out in inertial frames, some theorists have conjectured that GR might be an 'effective field theory'.[q]

It appears that one could interpret Einstein gravity as an effective gauge field theory to explain its consistency with experiments carried out in inertial frames. This may shed light on Wigner's comments: 'Expressing our results in terms of the values of coordinates became a habit with us to such a degree that we adhere to this habit also in general relativity, where values of coordinates are not per se meaningful.'[7]

To be an 'effective field theory' in inertial frames means that once the field equation of gravity is derived from the postulate of general coordinate invariance, one drops[r] the original postulate of 'general coordinate invariance' at the end of calculations. *The crucial step is that one simply identifies x^μ at the end of calculations in Einstein gravity as coordinates in an inertial frame so that numerical results can be compared with experiments, which are carried out in inertial frames.* In this sense, all numerical results of Einstein gravity effectively acquire a 'meaning' in inertial frames and hence, it can be tested by experiments carried out in inertial frames. In practice, this appears to be the way that GR has been used (or abused) to make predictions, such as the perihelion shift of Mercury, etc.

---

[q]For example, particle physicist T. D. Lee expressed such a view in conversations with colleagues at a High Energy Conference (1973, London). Similarly, E. C. G. Sudarshan in conversations with visitors and colleagues at the Center of Particle Theory, UT Austin (1972–1975) expressed a similar view.

[r]This resembles what Yukawa did in his formulation of non-local field theory to resolve the ultraviolet divergences in quantum field theories. At a Rochester conference in 1967, organized by R. E. Marshak, Yukawa responded to a question of consistency by Källén.[15]

Interestingly, a similar treatment of 'quantum Einstein gravity' was also carried out by Feynman and others. Around 1960, a formulation and a set of rules for quantum gravity based on Einstein's classical gravity was obtained by Feynman, DeWitt and Mandelstam (FDM).[16] Feynman ignored the procedure of field quantization in curved space-time and treated Einstein gravity as a perturbation theory in inertial frames. He was thus able to obtain the rules for quantum gravity with simple diagrams, just like in QED. Feynman claimed that by summing up all orders he had Einstein gravity. In this way, one may say that 'FDM quantum gravity' is an effective theory of quantum gravity, in which the coordinates x^μ are treated as if they were coordinates in inertial frames.

In light of previous discussions, one could interpret Einstein gravity as an 'effective gauge field theory.' Clearly, there is no logical connection between Einstein gravity and Yang-Mills gravity in flat space-time. However, this interpretation offers some sort of response to Wigner's question: "Most of us have struggled with the problem of how, under these premises, the general theory of relativity can make meaningful statements and predictions at all."[7] Although general relativity does not involve space and time coordinates or energy-momentum with operational meanings and energy is not conserved within the framework of general relativity, as implied by Noether's theorem II, it correctly predicts physical phenomena such as the perihelion motion of the Mercury, provided at the end of calculations, its coordinates x^μ are identified with those in an inertial frame. In this sense, general relativity is indeed a good 'effective gauge field theory.' Furthermore, in view of the formal similarities shown in Table A.1, it suggests that Yang-Mills gravity could play the role of this 'gauge field theory' in an inertial frame.

A rough comparison of quantum Yang-Mills (YM) gravity and quantum Feynman-DeWitt-Mandelstam (FDM) gravity, which is based on classical Einstein gravity, is shown in Table A.2 below.

Table A.2

Quantum YM Gravity	Quantum FDM Gravity
inertial frame	(*'effective inertial frame'*)
based on Yang-Mills gravity (translation T_4 gauge symmetry)	based on Einstein gravity (non-gauge-symmetry theory)

$$quantum\ tensor\ field:\ \phi_{\mu\nu}(x) \qquad \sqrt{G}h_{\mu\nu}(x) \equiv g_{\mu\nu} - \eta_{\mu\nu}$$

$$L_{YM} = \frac{1}{2g^2}(quadratic\ gauge\ curvature)^* \qquad L_{FDM} = \frac{1}{G}R\sqrt{-g}$$

gauge cov. deriv : $[\partial_\mu + g\phi_\mu^\nu(x)\partial_\nu]\psi$ *cov. deriv* : $\partial_\mu V^\nu(x) + \Gamma_{\mu\lambda}^\nu(x)V^\lambda(x)$

only attractive gravitational force between fermion-fermion and between fermion-antifermion (between all physical particles)   (no correspondence)

S matrix : $Feynman - Dyson\ rules$ (*S matrix* : *FD rules*)

$$\frac{-i}{2k^2}\left[\eta_{\alpha\beta}\eta_{\rho\sigma} - \eta_{\rho\alpha}\eta_{\sigma\beta} - \eta_{\rho\beta}\eta_{\sigma\alpha}\right]^{**} \qquad \propto \frac{1}{k^2}\left[\eta_{\mu\nu}\eta_{\sigma\tau} - \eta_{\mu\sigma}\eta_{\nu\tau} - \eta_{\mu\tau}\eta_{\nu\sigma}\right]$$

$$ghost\ propagator:\ \frac{-i}{k^2}\eta^{\mu\nu} \qquad \frac{-i}{k^2}\eta^{\mu\nu}$$

max.# of gravitons in a vertex : 4   *max.# of gravitons in a vertex* : ∞

non−renormalizable   *non−renormalizable*

$$non-dimensionless\ const:\ g = \sqrt{8\pi G} \qquad \sqrt{G}\left(= \sqrt{6.708 \times 10^{-39}\ GeV^{-2}}\right)$$

----

*: T_4 gauge curvature: $C^{\mu\nu\alpha} = J^{\mu\lambda}\partial_\lambda J^{\nu\alpha} - J^{\nu\lambda}\partial_\lambda J^{\mu\alpha}$, $\quad J^{\mu\lambda} = \eta^{\mu\lambda} + g\phi^{\mu\lambda}$.
**: Yang-Mills graviton propagator (in DeWitt gauge). The overall factor of $(1/2)$ in the propagator is necessary for Yang-Mills gravity to be consistent with its T_4 gauge identity — a generalized Ward-Takahasi identity for the Abelian T_4 group with ghosts.

References

1. J. P. Hsu and L. Hsu, *A Broader View of Relativity, General Implications of Lorentz and Poincaré Invariance* (2nd. ed., 2006, World Scientific) pp. 31–33 (Voigt), pp. 54–57; J. P. Hsu and Y. Z. Zhong, *Lorentz and Poincaré Invariance, 100 Years of Relativity* (World Scientific, 2001), pp. 76–146. A. Ernst and J. P. Hsu, 'First proposal of the universal speed of light by Voigt in 1887', Chinese J. Phys. 39 (3), 211–230 (2001).
2. D. Hilbert, 'The Foundations of Physics' in *100 Years of Gravity and Accelerated Frames, The Deepest Insights of Einstein and Yang-Mills* (Eds. J. P. Hsu and D. Fine), pp. 120–131.
3. E. Kretschman, Ann. Physik, **53** 575 (1917); P. Lenard, Uber das Rela. Ather, Grav. (Leipzig, 918; 2nd ed., 1920); Phys. Z. **21**, 666 (1920); see also E. P. Wigner in Ref. 7; W. Pauli, *Theory of Relativity* (tr. G. Field, Pergamon Press, London, 1958), p. 66, p. 149.
4. E. Noether, Goett. Nachr., 235 (1918). English translation of Noether's paper by M. A. Tavel is online. Google search: M. A. Tavel, Noether's paper.
5. L. Landau and E. Lifshitz, *The Classical Theory of Fields* (tr. M. Hamermesh, Addison-Wesley, 1951) p. 249, pp. 312–313, pp. 316–324.
6. V. Fock, *The Theory of Space Time and Gravitation* (tr. N. Kemmer, Pergamon Press, 1959) 163–166. See also Nobel Lecture, by Wigner in Ref. 7.
7. E. P. Wigner, *Symmetries and Reflections, Scientific Essays* (The MIT Press, 1967), pp. 52–53. See also Wigner in *NOBEL LECTURES, PHYSICS 1963–1970* (World Scientific, Singapore. New Jersey. London. Hong Kong, 1998) pp. 12–17.
8. S. S. Chern, *Lecture Notes on Differential Geometry* (Lian-Jing Publishing Co. Taipei, 1990, in Chinese.) p. iv, pp. 315–320. M. A. Melvin, Rev. Mod. Phys., **32**, 477 (1960).
9. F. J. Dyson, Bulletin of the American Math. Soc., 78, Sept. 1972. See also J. P. Hsu and D. Fine, *100 Years of Gravity and Accelerated Frames, The Deepest Insight of Einstein and Yang-Mills* (World Scientific, 2005) pp. 347–352, (with A Brief Remark for 'Missed Opportunity' by Dyson).
10. J. P. Hsu and L. Hsu, in *Space-Time, Yang-Mills Gravity, and Dynamics of Cosmic Expansion* (World Scientific, 2020), pp. 72–77, pp. 113–131, p. 145, pp. 232–239. See also M. Nowakowski, Mathematical Review Clippings (May 2021).
11. See Refs. 10, pp. 106–110.
12. S. M. Carroll, *Spacetime and Geometry (An Introduction to General Relativity)* (Addition Wesley, 2004), Appendix B.
13. See Ref. 8, p. 204.
14. W. Ślebodziński, Bull. Acad. Roy. Belg. **17**, 864 (1931).
15. H. Yukawa and G. Källén, in *Proc. of the 1967 International Conference on Particles and Fields* (Eds. C. R. Hagen, G. Guralnik and V. A. Mathur, Interscience, 1967.) p. 184.

16. J. P. Hsu and D. Fine, *100 Years of Gravity and Accelerated Frames, The Deepest Insight of Einstein and Yang-Mills* (World Scientific, 2005) pp. 272–324; S. Mandelstam, Phys. Rev. **175**, 1604 (1968); S. H. Kim and J. P. Hsu, Eur. Phys. J. Plus, **127**, 146 (2012).

Appendix B

A New Gauge Invariant Phase Equation for Bound Fermions in Superconductors

B-1. Introduction

The conventional treatment of Cooper pairs in superconductors and of the Josephson effect, which is based on the Schrödinger equation, is only approximately correct. We discuss a new superconducting phase equation, which reduces to the Schrödinger equation in the limit of zero phase (or when the Cooper pairs are broken). This new phase equation is consistent with the Meissner effect and flux quantization in a superconducting ring. We hope that the new phase equation will predict more accurately the effects of quantum Yang-Mills gravity on superconductors.

In the framework of field theories, gauge invariance has been established as the cornerstone for all interactions in physics. Here, we discuss physical effects of quantum Yang-Mills gravity on the gauge invariant phase equation for superconductors and on a Josephson junction. A new gauge-invariant equation for bound fermions intrinsically associated with an internal phase P is postulated and called the superconducting phase equation.

Internal gauge symmetries such as U_1 and SU_3 are violated by the presence of quantum Yang-Mills gravity with the universal gravitational interaction coupling to all non-gravitational fields.[1,2] This gravitational coupling leads to a difference of roughly 1 part in 10^9 in the Josephson effect. We discuss the feasibility of testing the physical effects of quantum Yang-Mills gravity using the superconductor voltage standard, which has a precision of the order of 10^{-10}.[3]

Bound fermions such as Cooper pairs in a superconductor are assumed to be intrinsically associated with a 'phase field' $P(x) \equiv P(\mathbf{r}, t)$, which appears as a phase in the 'boson wave function' $B(x)$ of bound fermions. The phase P does not have a wave equation by itself in general. $P(x)$ may be interpreted as an 'internal potential field' associated with bound fermions, and it cannot exist when the bond is broken. If the scalar phase $P(x)$ vanishes, the phase equation has no physical meaning for superconductors, although it formally reduces to the form of the Schrödinger equation for a quantum 'particle' with charge $2e$ and mass $2m_e$ (which does not exist as a physical particle in vacuum).

The superconducting (SC) phase equation involving space and time derivatives of the phase P gives a coherent explanation of flux quantization, the Meissner effect and the Josephson effect. Thus, the SC phase equation is postulated to be the basic 'quantum equation' for the dynamics of superconductivity.

In comparison with normal electrons, bound pairs are postulated to have a new 'physical' phase $P(x)$. Such a phase P by itself is not observable and is not a usual field generated by a source. It is intrinsically internal and distinct from all other known fields in quantum mechanics and field theory. It appears that the SC phase $P(x)$ has physical consequences only in the presence of other external fields such as the electromagnetic fields. That is, the time derivative of the phase $\hbar \partial P / \partial t$ could play the role of a voltage and the space derivative of the phase $\hbar \partial_k P$ could cause a supercurrent.

The basis of superconductivity is a small net effective attraction between electrons in matter, so that they form bound pairs with new physical properties. This attraction is due to the interactions of the electrons with the vibrations of the atoms in the lattice. For a basic physical phenomenon, there must be an equation. These bound states of pairs are postulated to be described by a gauge invariant phase equation for a physical system at essentially zero temperature. Moreover, the boson wave function $B(x)$ actually describes billions of pairs in the same state rather than just one pair.[4]

B-2. Gauge invariant superconducting phase equation

Based on a phase gauge symmetry, we postulate the pure phase equation,

$$\left(i\hbar \frac{\partial}{\partial t} + \hbar \frac{\partial P}{\partial t} \right) B(x) = \frac{1}{2M} (-i\hbar \partial_k - \hbar \partial_k P)^2 B(x), \qquad (B1)$$

which is invariant under a new U_1 phase transformation with the 'phase gauge function' Λ_p,

$$B'(x) = B(x) exp[i\Lambda_p(x)], \quad P'(x) = P(x) + \Lambda_p(x), \qquad (B2)$$

where $(x) = (\mathbf{r}, t)$.

In the presence of electromagnetic potentials, we generalize (B1) to the superconducting (SC) phase equation for bound electrons or Cooper pairs with arbitrary (A_0, A_k):

$$\left(i\hbar \partial_t - QA_0 + \hbar \partial_t P + Q \left[\int_s^{\mathbf{r}} \partial_t \mathbf{A}(\mathbf{r}', t) \cdot d\mathbf{r}' \right]_{Le} \right) B \qquad (B3)$$

$$= \frac{1}{2M} \left[-i\hbar\partial_k - QA_k - \hbar\partial_k P + Q \int_{t_o}^{t} \partial_k A_0(\mathbf{r}, t') dt' \right]^2 B,$$

$$A_0 = A_0(\mathbf{r}, t), \quad A_k = A_k(\mathbf{r}, t), \quad Q = 2e, \quad M = 2m_e,$$

$$\partial_k = (\partial_x, \partial_y, \partial_z), \quad \partial_t = \partial/\partial t,$$

where the term involving a 3-dimensional integral $\int^{\mathbf{r}}$ in (B3) should be understood as Hamilton's characteristic function, to be explained below (after Eq. (B10)).

The general SC phase Eq. (B3) is invariant under a generalized electromagnetic gauge transformation involving two arbitrary gauge functions, the usual U_1 electromagnetic gauge function $\Lambda_e(x)$ and a new U_1 phase gauge function $\Lambda_p(x)$:

$$B'(x) = B(x) exp[-2iQ\Lambda_e(x)/\hbar + i\Lambda_p(x)], \tag{B4}$$

$$A_0'(x) = A_0(x) + \partial_t \Lambda_e(x), \tag{B5}$$

$$A_k'(x) = A_k(x) - \partial_k \Lambda_e(x), \tag{B6}$$

$$P'(x) = P(x) + \Lambda_p(x), \quad x = (\mathbf{r}, t). \tag{B7}$$

Since $\Lambda_p(x)$ and $\Lambda_e(x)$ are two independent arbitrary gauge functions of space-time x, the present dynamical model of superconductivity has dual gauge symmetries, namely the 'electromagnetic' and 'phase' gauge symmetries. The SC phase Eq. (B3) is invariant under the $U_1^{em} \times U_1^p$ gauge transformations (B4)–(B7). This dual gauge symmetry for the dynamics of superconductivity is a generalization of the usual U_1 gauge symmetry of electrodynamics.

Suppose one defines $\Lambda_p(x) = 2Q\Lambda_e(x)/\hbar$, the transformations (B4)–(B7) reduce to

$$B'(x) = B(x), \quad A_0'(x) = A_0(x) + \partial_t \Lambda_e, \tag{B8}$$

$$A_k'(x) = A_k(x) - \partial_k \Lambda_e, \quad P'(x) = P(x) + 2Q\Lambda_e/\hbar,$$

which resemble the conventional 'gauge transformations' in the discussions of the Josephson junction.[1] This type of special gauge transformation is assumed in the conventional discussions of Josephson junctions in order to obtain the Josephson current-phase relation in the presence of electromagnetic fields, in which the wave function is unchanged, i.e., is not involved in the gauge transformations.[1] We note that such a gauge invariance could not be applied in a Lagrangian formulation for superconductivity because the gauge invariant Lagrangian of fields must involve the wave function in the gauge transformations.

If $A_0 = 0$ and $\mathbf{A} = 0$, the SC phase Eq. (B3) reduces to the pure phase Eq. (B1). If one were to write the boson wave function in the form, $B = B_d(x)exp[iP(x)]$, one would have formally the 'free' equation for $B_d(x)$

$$i\hbar\partial_t B_d = \frac{1}{2M}\left[-i\hbar\partial_k\right]^2 B_d. \tag{B9}$$

The concept of phase plays a special role in superconductivity. Physically, the reason to postulate the SC phase Eq. (B3) to have the two terms with $d\mathbf{r}'$ (i.e., involving Hamilton's characteristic function) and dt is precisely to have the following new phase in the wave function $B(x) = B(\mathbf{r}, t)$,

$$B(\mathbf{r}, t) = G_d(\mathbf{r}, t)exp\left[+iP(\mathbf{r}, t) - \frac{iQ}{\hbar}\int_{t_o}^t A_0(\mathbf{r}, t')dt' \right.$$

$$\left. + \frac{iQ}{\hbar}\left(\int_{\mathbf{s}}^{\mathbf{r}} \mathbf{A}(\mathbf{r}', t)\cdot d\mathbf{r}'\right)_{Le} \right], \tag{B10}$$

$$G_d(\mathbf{r}, t) = G(\mathbf{r})exp(-iEt/\hbar),$$

where the initial points $\mathbf{s}$ and t_o are arbitrarily fixed points and the end points $\mathbf{r}$ and t of the integrals are variable. Therefore, the physical effects of the general electromagnetic potentials A_0, A_k and the phase P in superconductors are completely dictated by the phase of the boson wave function $B(\mathbf{r}, t)$ in (B10).

In order to ensure the local properties of all terms in the SC phase Eq. (B3) and in the solution $B(\mathbf{r}, t)$ in (B10), we must consider the path integral $\left(\int_{\mathbf{s}}^{\mathbf{r}} \mathbf{A}(\mathbf{r}', t)\cdot d\mathbf{r}'\right)_{Le}$ to be an action integral, i.e., Hamilton's characteristic function.[2,5,6] The starting point s is fixed, the end point is variable. Only the actual paths are allowed, which are denoted by $()_{Le}$. The subscript in $()_{Le}$ means that the path from initial fixed point s to the variable end point $\mathbf{r}$ must satisfy the Lagrangian equation derived from the action integral H_a.[3] Thus, we have the local relation

$$\partial_k H_a(\mathbf{r}, t) = A_k(\mathbf{r}, t), \qquad H_a \equiv \left(\int_{s_k}^{r'_k=r_k} A_i(\mathbf{r}', t)dr'_i\right)_{Le}$$

$$\delta H_a = A_i(\mathbf{r}',t)\delta r_i'|_{s_k}^{r_k} + \int_{s_k}^{r_k}\left(\frac{\partial A_i}{\partial r_k'}dr_i' - dA_k\right)\delta r_k', \qquad \text{(B11)}$$

where the path in (B11) is required to satisfy[6] the Lagrange equation,

$$\frac{\partial L}{\partial r_k'} - \frac{d}{dt}\frac{\partial L}{\partial \dot{r}_k'} = 0, \qquad H_a = \int L\,dt, \quad L = A_i(r_k',t)\dot{r}_i',$$

or, equivalently,

$$\frac{\partial A_i}{\partial r_k'}dr_i' - dA_k(\mathbf{r}',t) = 0.$$

This relation holds automatically for the one-dimensional case. Thus, the function H_a in (B11) is just an ordinary indefinite integral for the one-dimensional case.

Since electron pairs are bosons, when there are many of them in a given state in a superconductor, there is an especially large amplitude for other pairs to go to the same state.[4] The bound-fermion-phase current density j_k is defined to satisfy the continuity equation

$$\partial_t\rho = -\partial_k j_k, \qquad \rho = QB^*B. \qquad \text{(B12)}$$

Thus, we derive the current density j_k from the SC phase Eqs. (B3) and (B12),

$$j_k = \frac{Q}{2M}\left[B^*\left(-i\hbar\partial_k + Z_k + Q\int^t \partial_k A_0(\mathbf{r},t')dt'\right)B\right.$$

$$\left. + B\left(-i\hbar\partial_k + Z_k + Q\int^t \partial_k A_0(\mathbf{r},t')dt'\right)^* B^*\right] \qquad \text{(B13)}$$

$$Z_k \equiv -QA_k - \hbar\partial_k P.$$

In the dynamic model of superconductivity based on the SC phase Eq. (B3), the density $\rho(x) = QB^*(x)B(x)$ in (B12) does not describe the usual quantum mechanical probability density of one Cooper pair. Rather, $\rho(x) = QB^*(x)B(x)$ is postulated to be interpreted as the macroscopic charge density of billions of bound-electron bosons in the same state at essentially zero temperature. In this case, the density $\rho(x)$ is almost a perfect constant.[4]

B-3.   Quantum Yang-Mills gravity and the superconducting phase equation

In order to discuss an experimental test of quantum Yang-Mills gravity using the Josephson effect in a superconductor, let us consider the modification of the dynamic Eq. (B3) by gravity. In the presence of Yang-Mills gravity with the universal gravitational coupling based on space-time translational gauge symmetry, the ordinary derivatives are assumed to be replaced by the T_4 translational gauge covariant derivatives,

$$\partial_\mu \to \partial_\mu - i(g/\hbar)\phi_\mu^\nu p_\nu = (\delta_\mu^\nu + g\phi_\mu^\nu)\partial_\nu \equiv J_\mu^\nu \partial_\nu, \qquad \text{(B14)}$$

$$p_\nu = i\hbar\partial_\nu, \quad J_\mu^\nu = \delta_\mu^\nu + g\phi_\mu^\nu = J_{\mu\alpha}\eta^{\alpha\nu},$$

$$\eta^{\alpha\nu} = (1,-1,-1,-1), \qquad \mu,\nu,\alpha = 0,1,2,3,$$

in the Lagrangians (and equations) of all non-gravitational fields. On the surface of the Earth, with $r = R_E$ and mass m_E, the non-vanishing components J_μ^μ (to the first-order approximation) are[1]

$$J_0^0 \approx 1 + g\phi_0^0 \equiv h_0, \quad J_1^1 = J_2^2 = J_3^3 \approx 1 - g\phi_0^0 \equiv h_1; \qquad \text{(B15)}$$

$$g\phi_0^0 = \frac{Gm_E}{c^2 R_E} \approx 6.95 \times 10^{-10}, \qquad G = g^2/(8\pi c^3).$$

The dimensionless Earth surface potential, $g\phi_0^0 = Gm_E/c^2 R_E$ can be approximated as a constant. Thus, in the presence of gravity, the SC phase Eq. (B3) is modified by the replacements,

$$\partial_t \to h_0\partial_t, \qquad \partial_k \to h_1\partial_k, \qquad B \to B_g, \qquad \text{(B16)}$$

to the lowest order approximations in g. Accordingly, the SC phase Eq. (B3) and the current density j_k in (B13) modified by gravity are respectively given by

$$\left[i\hbar h_0\partial_t - QA_0 + \hbar h_0\partial_t P + Q\left(\int^{\mathbf{r}} h_0\partial_t\mathbf{A}(\mathbf{r}',t)\cdot d\mathbf{r}'\right)_{Le}\right]B_g$$

$$= \frac{1}{2M}\left[-i\hbar h_1\partial_k - QA_k - \hbar h_1\partial_k P + Q\int^t h_1\partial_k A_0(\mathbf{r},t')dt'\right]^2 B_g, \qquad \text{(B17)}$$

and

$$j_{gk} = \frac{Q}{2M}\left[B_g^*\left(-i\hbar h_1\partial_k + Y_k + Q\int^t h_1\partial_k A_0(\mathbf{r},t')dt'\right)B_g\right.$$

$$\left. +B\left(-i\hbar h_1\partial_k + Y_k + Q\int^t \partial_k h_1 A_0(\mathbf{r},t')dt'\right)^* B_g^*\right], \qquad \text{(B18)}$$

$$Y_k \equiv -QA_k - \hbar h_1\partial_k P, \quad \rho_g = QB_g^* B_g, \quad h_0\partial_t\rho_g = -h_1\partial_k j_{gk}.$$

We note that the modified SC phase Eq. (B17) is still formally invariant under the modified gauge transformations (B4)–(B7) with the replacements (B16). However, the gauge symmetries with internal gauge groups are in general, violated by quantum Yang-Mills gravity.[1]

B-4. Modified superconducting phase equation and the Josephson relations

Let us analyze a Josephson effect based on the modified SC phase Eq. (B17) and compare it with the usual Josephson relations. We consider the solution of the SC phase Eq. (B17) in the presence of a tunneling potential V_o, making the replacement

$$-QA_0 \to -QA_0 - V_o,$$

in Eq. (B17) to include a tunneling potential V_o in a Josephson junction.

For simplicity and without loss of generality of the gravitational effect, let us consider the one-dimensional case, $\mathbf{r} \to z$, in (B10) for general electromagnetic potentials (A_0, A_z). The static wave function $G(\mathbf{r}) = G(z)$ in (B10) in the insulating region satisfies[6]

$$-\frac{\hbar^2}{2M}h_1^2\partial_z^2 G(z) = (h_0 E - V_o)G(z), \quad |z| \le a, \tag{B19}$$

where $V_o > E$ and we have used the replacements (B16). Equation (B19) leads to a simple solution,

$$G(z) = G_1 cosh(z/z_c) + G_2 sinh(z/z_c), \tag{B20}$$

$$z_c = \frac{\hbar h_1}{\sqrt{2M(V_o - h_0 E_o)}},$$

in the insulating region $|z| \le a$.

Since we are interested in the effects due to the presence of Yang-Mills gravity and an arbitrary electromagnetic potential (A_0, A_k), it suffices to have solutions for B_g for the one-dimensional case $(\mathbf{r} \to z)$. From (B10), (B16) and (B20), we have the solution for the boson wave function B_g,

$$B_g(z,t) = [G_1 cosh(z/z_c) + G_2 sinh(z/z_c)]e^{iX}, \tag{B21}$$

$$X = -Et/\hbar + P + \frac{Q}{\hbar}\int^z A_z(z,t)dz - \frac{Q}{\hbar}\int^t A_0(z,t)dt.$$

To determine the parameters G_1 and G_2 in (B20), we impose the boundary conditions at $z = \pm a$ for $B_g(z,t)$ in (B21). The boundary conditions at $z = -a$ and $z = a$ for B_g are required to have electromagnetic gauge invariant phases ϕ_2 and ϕ_1,

$$B_g(-a,t) = \sqrt{\rho_o}\, e^{i\phi_1}, \qquad B_g(a,t) = \sqrt{\rho_o} e^{i\phi_2}, \qquad \text{(B22)}$$

$$\phi_1 = \phi(-a,t) = -\frac{Q}{\hbar}\left[\int^t A_0(-a,t')dt' + \int^{-a} A_z(z,t)dz\right], \qquad \text{(B23)}$$

$$\phi_2 = \phi(a,t) = -\frac{Q}{\hbar}\left[\int^t A_0(a,t')dt' + \int^a A_z(z,t)dz\right], \qquad \text{(B24)}$$

where ρ_o is the constant charge density at $z = -a$ and $z = +a$.

To find the solution in the insulating region for general fields (A_0, A_z), we use the solution (B21) and the boundary conditions in (B22) to determine for the coefficients G_1 and G_2,

$$G_1 = \frac{\sqrt{\rho_o}[exp(-iZ_1) + exp(-iZ_2)]}{2cosh(a/z_c)}, \quad Z_1 = \phi_1 - X_1, \qquad \text{(B25)}$$

$$G_2 = \frac{\sqrt{\rho_o}[-exp(-iZ_1) + exp(-iZ_2)]}{2sinh(a/z_c)}, \quad Z_2 = \phi_2 - X_2, \qquad \text{(B26)}$$

$$Z_1 = Z(-a,t), \quad Z_2 = Z(a,t), \quad X_1 = X(-a,t), \quad X_2 = X(a,t).$$

Note that in a superconducting material, there is a background of positive charge due to the atomic ions of the lattice. Because there is no net charge, the charge density ρ_o is almost perfectly uniform[4] so that the boundary conditions in (B22) can be imposed.

For the 1-dimensional case in the insulating region, we have the current density $j_{gk} = j_{gz} = j_g$ (in the presence of gravity), where j_{gk} is given in (B18). At the boundaries of the electrodes $z = \pm a$, we have the current density

$$j_g = \frac{Qh_1\hbar}{2M}\left(B_g^*[-i\partial_z - \frac{Q}{h_1\hbar}A_z - \partial_z P + \frac{Q}{\hbar}\int^t \partial_z A_0(z,t')dt']B_g + c.c.\right),$$

$$= \frac{Qh_1\hbar}{M}Im(G^*(z)\partial_z G(z)) = \frac{Qh_1\hbar}{Mz_c}Im(G_1^* G_2), \qquad \text{(B27)}$$

where we have used (B10), (B13), (B16) and c.c. denotes the complex conjugate. Note that since there are no derivatives in (B10), B_g is formally the same as B. Based on (B25)–(B27), we obtain

$$j_g = j_{gc} sin(P_{g21}), \quad j_{gc} = \frac{Q\hbar\rho_o}{Mz_c sinh(2a/z_c)}, \tag{B28}$$

where z_c is given in (B20). In the presence of gravity, the phase difference P_{g21} between the two points $z = \pm a$ is given by

$$P_{g21} = \phi_2 - X_2 - \phi_1 + X_1, \tag{B29}$$

$$\phi_2 - \phi_1 = -\frac{Q}{\hbar}\left[\int^t [A_0(a,t) - A_0(-a,t)]dt + \int_1^2 A_z(z,t)dz\right], \quad \int_1^2 \equiv \int_{-a}^a,$$

$$X_2 - X_1 = P(a) - P(-a) + \frac{Q}{\hbar}\int_{-a}^a A_z(z,t)dz - \frac{Q}{\hbar}\int^t [A_0(a,t) - A_0(-a,t)]dt.$$

Equations (B28) and (B29) are the Josephson current-phase relations in the presence of Yang-Mills gravity. The time derivative of the phase difference P_{g21} in (B29) in the presence of gravity is

$$h_0\partial_t P_{g21} = \frac{Qh_0}{\hbar}\int_1^2 E_z dz + (h_0 - 1)\frac{Q}{\hbar}\int_1^2 [\partial_z A_0(z,t)]dz, \tag{B30}$$

where we have used the following relations,

$$h_0\partial_t(\phi_2 - \phi_1) = -\frac{Qh_0}{\hbar}\left[\int_1^2 [\partial_z A_0(z,t) + \partial_t A_z(z,t)]dz\right] \tag{B31}$$

$$= \frac{Qh_0}{\hbar}\int_1^2 E_z dz,$$

$$h_0\partial_t P = \frac{Q}{\hbar}A_0 - \frac{Q}{\hbar}\int^z h_0\partial_t A_z(z',t)dz' + \frac{M}{\hbar 2\rho_g^2}j_{gz}^2, \tag{B32}$$

where

$$j_{gz} = (\rho_g/M)[-QA_z - \hbar h_1\partial_z P + Q\int^t h_1\partial_z A_0 dt].$$

The time derivative of the phase $\partial_t P$ and the current density j_{gz} in (B32) at the boundaries $z = \pm a$ are obtained from (B17) for the one-dimensional case with the assumption $B_g = constant$. Using the expression $X_2 - X_1$ in (B29) and $j_{gz}^2(-a,t) = j_{gz}^2(a,t)$ in (B32),[6] we obtain

$$-h_0\partial_t(X_2 - X_1) = (h_0 - 1)\frac{Q}{\hbar}\int_1^2 [\partial_z A_0(z,t)]dz. \tag{B33}$$

The first term on the right side of (B30) in the absence of gravity (i.e., $h_0 = 1$) is formally the same as that in the Josephson current-phase relation. However, there is a difference in their physical origin. Namely, Josephson's result for the time rate-of-change of the phase difference is not derived from a gauge invariant phase equation for superconductors in the presence of general electromagnetic potentials (A_0, A_k). Rather, the usual result is first obtained for the special case $(A_0 = 0, A_k = 0)$. Then, in the presence of general electromagnetic potentials (A_0, A_k), a 'gauge invariant' phase difference is defined (or 'postulated') to obtain the conventional Josephson current-phase relation.[7] We note that the 'gauge invariant phase' in this case is not the usual electromagnetic U_1 gauge invariance because

(i) the wave function is not involved in the gauge transformations,

(ii) a new function θ is involved in the gauge transformation, and

(iii) the scalar potential is specifically chosen to be not involved in the transformations.

In general, if the electromagnetic scalar potential is involved in the transformations, then there will be an extra parameter involved in the Josephson's gauge invariant phase difference. In contrast, the present model is based on the SC phase Eq. (B3), which is invariant under both the conventional electromagnetic gauge transformations and the phase gauge transformations (involving a new gauge function Λ_p), as given in (B4)–(B7).

The second term in (B30) is very small because $(h_0 - 1) \approx g\phi_0^0 \approx 10^{-9}$ and violates U_1 gauge invariance. Such a violation is due to the presence of Yang-Mills gravity.[1]

B-5.   Tests of quantum Yang-Mills gravity with superconductors

B-5.1.   *Modified Josephson current-phase relations*

We note that the modifications to the voltage $h_0 \partial_t P_{g21} = V_{g21}$ in (B30) are due to quantum Yang-Mills gravity rather than classical Yang-Mills gravity. The T_4 (space-time translation) gauge covariant derivatives in (B15) are related to quantum fields in a Lagrangian. Classical gravity is based on the Hamilton-Jacobi type equation[8] for the motion of macroscopic objects and light rays. Our discussions here are not related to the classical Hamilton-Jacobi equation in Yang-Mills gravity. The gravitational effects in the previous discussions for (B14) to (B34) are based on the space-time translational gauge covariant derivatives in quantum Yang-Mills gravity.[1]

Equations (B28) and (B29) are the modified Josephson current-phase relations in the presence of Yang-Mills gravity. To facilitate comparison with experimental measurements, we can express the time derivative of the phase difference (B30) in a slightly different form by using $h_0 = 1 + g\phi_0^0$. One term is independent of gravity, and the other has a very small dependence on gravity $g\phi_0^0$,

$$V_{g21} = h_0 \partial_t P_{g21} = \frac{Q}{\hbar} \left[\int_1^2 E_z dz - g\phi_0^0 \int_1^2 \partial_t A_z(z,t)dz \right]. \qquad (B34)$$

Based on the new SC phase Eq. (B17), we obtain the physical effects (B34) of quantum Yang-Mills gravity on a Josephson junction in the presence of an arbitrary electromagnetic potential (A_0, A_k). The resultant voltage in (B34), i.e., $V_{g21} = h_0 \partial_t P_{g21}$ can be tested experimentally.

For simplicity, we consider the special case $E_z = 0$ in (B34). In this case, the voltage difference modified by gravity depends only on the vector potential $A_z(z,t)$ and the dimensionless potential $g\phi_0^0$ on the surface of the Earth. The SC phase Eq. (B17) and Yang-Mills gravity predicts a new very small voltage difference V'_{g21}:

$$V'_{g21} = -g\phi_0^0 \left(\frac{Q}{\hbar} \int_1^2 \partial_t A_z(z,t)dz \right), \qquad (B35)$$

where $g\phi_0^0 \approx 6.95 \times 10^{-10}$ is the dimensionless potential at the surface of the Earth.

One can use a strong time-varying magnetic field, $\mathbf{B} = \nabla \times \mathbf{A}$, to produce a large time-dependent $\partial_t A_z(z,t)$. In this case, the small voltage V'_{g21} in (B35) could be detectable as the precision in the Josephson junction voltage standard[3] is on the order of 10^{-10}. The prediction (B35) can be tested by comparing the voltage across a Josephson junction in a laboratory at rest on Earth with that across a junction in free fall (e.g., in the International Space Station).

B-5.2. *Flux quantization in a superconducting ring*

Flux quantization in a superconducting ring is also modified by quantum Yang-Mills gravity. In this case, the modified charge density $QB_g^*B_g = \rho_g$ is constant in the superconductor. Also, well inside the body of the ring, the current density in (B32) vanishes.[4] In the 3-dimensional case, $j_{gk} = 0$, we have,

$$\hbar h_1 \nabla \left(P - \frac{Q}{\hbar} A_{I0} \right) = -Q\mathbf{A}, \qquad A_{I0} = \int^t A_0 dt. \qquad (B36)$$

We can impose the usual gauge condition $\boldsymbol{\nabla} \cdot \mathbf{A} = 0$ in (B36), so that we have the following relation

$$(\boldsymbol{\nabla})^2 \left(P - \frac{Q}{\hbar} A_{I0} \right) = 0 \tag{B37}$$

everywhere inside the superconductor. This property (B37) implies that

$$\left(P - \frac{Q}{\hbar} A_{I0} \right) = constant. \tag{B38}$$

Consider the line integral of $\mathbf{A}$ in (B36) around a curve that goes around a physical ring near the center of its cross-section so that it never gets near the surface.[4] The line integral of the vector potential $\mathbf{A}$ around any loop is equal to the flux Φ_f of the magnetic field $\mathbf{B}$ through the loop: $\oint \mathbf{A} \cdot d\mathbf{r} = \Phi_f$. Equation (B36) leads to

$$\oint \boldsymbol{\nabla} \left(P - \frac{Q}{\hbar} A_{I0} \right) \cdot d\mathbf{r} = -\frac{Q}{\hbar h_1} \Phi_f. \tag{B39}$$

We note that although the loop integral on the left side of (B39) vanishes in a simply-connected superconductor, that is not necessary true for the ring-shaped superconductor under consideration here.[4] The expression $[P - (Q/\hbar)A_{I0}]$ is a phase of the wave function B_g in Eq. (B10) because we have the relation,

$$i \left[P - \frac{Q}{\hbar} A_{I0} \right] = iP(\mathbf{r}, t) - \frac{iQ}{\hbar} \int_{t_o}^{t} A_0(\mathbf{r}, t') dt'. \tag{B40}$$

We can require that the wave function is single valued, so that the loop integral of $\boldsymbol{\nabla}(P - (Q/\hbar)A_{I0})$ in (39) must be $2\pi n$, where n is an integer. Thus, we have a modified quantized flux Φ_f in the presence of Yang-Mills gravity in an Earth laboratory,

$$\Phi_f = \frac{h_1 \hbar \pi n}{q_e}, \qquad h_1 \approx (1 - 7 \times 10^{-10}), \qquad q_e > 0, \tag{B41}$$

where $Q = -2q_e$. Evidently, the basic flux unit $\Phi_{f0} = h_1 \hbar \pi / q_e$ is modified by Yang-Mills gravity by the factor h_1.

B-5.3.   *Modified Meissner effect*

In order to explain Meissner's observation that the magnetic field is expelled from a superconductor, London and London originally proposed the equation $j_k = -\lambda^2 A_k$, where j_k is the source of the vector potential A_k.[8] In the presence of Yang-Mills gravity, it is natural and logical to derive the equation from the Lagrangian L_{sc} for the SC phase field B_g and the electromagnetic potentials $A_\mu = (A_0, A_k)$,

$$L_{sc} = B_g^* \left[i\hbar h_0 \partial_t - Q' A_0 + \hbar h_0 \partial_t P + Q' \left(\int^{\mathbf{r}} h_0 \partial_t \mathbf{A}(\mathbf{r}', t) \cdot d\mathbf{r}' \right)_{Le} \right] B_g$$

$$+ \frac{1}{2M} \left[-i\hbar h_1 \partial_k - Q' A_k - \hbar h_1 \partial_k P + Q' \int^t h_1 \partial_k A_0(\mathbf{r}, t') dt' \right]^* B_g^* \quad \text{(B42)}$$

$$\times \left[-i\hbar h_1 \partial_k - Q' A_k - \hbar h_1 \partial_k P + Q' \int^t h_1 \partial_k A_0(\mathbf{r}, t') dt' \right] B_g$$

$$- \frac{1}{16\pi} (D_\mu A_\nu - D_\nu A_\mu)(D^\mu A^\nu - D^\nu A^\mu), \quad \eta_{\mu\nu} = (1, -1, -1, -1),$$

where $\mu, \nu = 0, 1, 2, 3$, and $D_\mu = \partial_\mu + g\phi_\mu^\nu \partial_\nu$ is the 4-dimensional translation gauge covariant derivative.[1] The Lagrangian L_{sc} contains both 3-dimensional terms associated with the SC phase equation with B_g and $Q' = Q/c$, and 4-dimensional terms for the electromagnetic potentials (A_0, A_k) in Gaussian (cgs) units. In order to derive the Meissner effect, it suffices to consider the vector potential A_k in the static case.

From the Lagrange equation $\partial_\alpha[\partial L/\partial(\partial_\alpha A_\beta)] - \partial L/\partial A_\beta = 0$, one can see that the Cooper-pair current density of the phase wave function B_g automatically plays the role of the source j_k of the vector potential A_k and leads to the Meissner effect. To a first-order approximation, the non-zero components of $g\phi_\mu^\nu$ are constant, so that we have the relation $\partial_\nu D^\beta = D^\beta \partial_\nu$. Thus, we can impose the usual Lorentz gauge condition $\partial_\mu A^\mu = 0$ to simplify the equation. Since the Meissner effect involves only the magnetic field or the vector potential $\mathbf{A}$, we may set $A_0 = 0$ and concentrate on the vector potential A_k,

$$-(c^{-2}\partial_t^2 - \partial_j^2)A_k + (4\pi Q/Mc^2)B_g^*(-QA_k)B_g = 0, \quad \text{(B43)}$$

where we have used $B_g = constant$ for superconductors in the static case. In this case, it suffices to consider the steady situation or the static limit related to the vector potential A_k,

$$\nabla^2 A_k - \lambda_o^2 A_k = 0, \quad \lambda_o^2 = 4\pi Q^2 B_g^* B_g/(Mc^2). \quad \text{(B44)}$$

To see the qualitative properties of the solution, we consider the 1-dimensional solution of (B44). One has a solution of the form $e^{-\lambda_o z}$, which implies that the potential field must decrease exponentially from the surface into the superconductor. Thus, the vector potential field only penetrates a thin layer (about $1/\lambda_o$) at the surface. One can use spherical coordinates to find a solution to (B44). The solution has the form of the Yukawa potential $A(r) = e^{-\lambda_o r}/r$.[10] It is estimated that $1/\lambda_o \approx 10^{-6} cm$.[4, 10]

B-6. Summary

A new superconducting (SC) phase Eq. (B3) is postulated for arbitrary electromagnetic potentials and an internal phase P for Cooper pairs. It is invariant under the $U_1^{em} \times U_1^p$ gauge transformations (B4)–(B7). The phase equation is generalized to (B17), which includes the presence of quantum Yang-Mills gravity and the universal gravitational interaction.[11] Similar to standard gauge theories, Yang-Mills gravity is based on space-time translational gauge symmetry. The generators of the group are the momentum operators and the spin-2 field appears naturally in this context.[12] Classical results are encoded in what we call an Einstein-Grossmann equation, which is used to compare the prediction of the theory with observation. Yang-Mills gravity brings gravity back into the arena of gauge fields in flat space-time and is consistent with experiments.

The phase equation leads to a modified Josephson current-phase relation, flux quantization, and Meissner effect. These modifications differ from the usual predictions by about 10^{-9} on the surface of the Earth. The proposed experiment with the Josephson effect is particularly interesting because it appears to be feasible, in contrast to the modifications in flux quantization and the Meissner effect in (B41) and (B44). The Josephson effect actually could test two physical effects of quantum Yang-Mills gravity:

(i) The SC phase equation leads to (B34) and (B35), which predict that in the presence of gravity, the phase-voltage relation in a Josephson junction is modified by the factor $g\phi_0^0$. Result (B35) can be tested by comparing the voltage across a Josephson junction in a laboratory at rest on Earth with that across a junction in free fall (e.g., in the International Space Station or in a plane maneuvering to simulate zero-gravity such as NASA's now-retired "Vomit Comet"). The prediction (B35) with $g\phi_0^0 \approx 10^{-9}$ on the Earth could be tested because the voltage standard with Josephson effects has a precision of $\approx 10^{-10}$.[3]

(ii) The modified voltage $V_{g21} = h_0 \partial_t P_{g21}$ given by (B34) in the presence of quantum Yang-Mills gravity is not electromagnetic U_1 gauge invariant. Only in the absence of gravity, i.e., $g\phi_0^0 = 0$ or $h_0 = 1$, is the voltage V_{g21} in (B34) gauge invariant. Thus, an experimental test of (B35) also provides a test of the violation of the U_1 gauge symmetry predicted by quantum Yang-Mills gravity.

It is interesting and potentially significant to test whether the electromagnetic gauge symmetry and the associated conservation of the electric charges are absolute in nature. We hope that the experiment with Josephson effects to test quantum Yang-Mills gravity can be performed in the near future.

References

1. J. P. Hsu and L. Hsu *"Space-Time, Yang-Mills Gravity and Dynamics of Cosmic Expansion"*, (World Scientific, 2019) pp. 202–204.
2. J. P. Hsu, Chin. Phys. C, **41**, 015101 (2017).
3. M. Grundmann, 'Dimensionality' in Encyclopedia of Condensed Matter Physics, 2005. Josephson Junctions. (accessed 17 Jan. 2022) 'A Josephson junction standard can realize a voltage with an accuracy of 10^{-10}.'
4. R. Feynman, 'A Seminar on Superconductivity' in The Feynman Lecture on Physics, vol. III, Ch. 21.
5. W. Yourgrau and S. Mandelstam, *Variational Principles in Dynamics and Quantum Theory*, Dover, 3rd ed, (1970) p. 50.
6. L. Landau and E. Lifshitz, *The Classical Theory of Fields*, Addison-Wesley (1951), p. 29.
7. http://www.nobel.se/physics/laureates/1973/giaever-lecture.pdf. Josephson Junction. Massachusetts Institute of Technology. 6.763 2003 Lecture 11. (accessed 3 Aug. 2022).
8. L. Hsu and J. P. Hsu, Chinese Phys. C, **43** (10), 105103 (2019). arXiv: 1908.01585 [physics.gen-ph].
9. F. London and H. London, Proc. Roy. Soc. (London) A 149, 71 (1935).
10. J. D. Jackson, *Classical Eectrodynamics* (3rd ed. Wiley, 1998) p. 600.
11. J. P. Hsu and S. H. Kim, Eur. Phys. J. Plus, **127**, 146 (2012).
12. M. Nowakowski, Mathematical Review Clippings, May 2021. MR3971218.

Appendix C

Quantum Shell Model for 3-Quark Bound States

C-1. Quantum 3-body problem: harmonic oscillator potential for a color-charge shell in baryons

In this section, we explore the dynamics of the first 3-quark confinement processes that produced baryons and anti-baryons soon after the birth of the universe.[1]

We consider a greatly simplified model for relativistic bound states of 3-quark baryons.[2] Baryons are typically supposed to be formed in a 3-body collision of quarks. However, such collisions presumably have an extremely small probability of occurrence and there is no clear picture for the configuration of a 3-quark bound state. However, these difficulties could be surmounted by a confining 3-quark (C3Q) model based on the idea that a relativistic 3-quark state can be formed by consecutive 2-body collisions.

For example, consider a proton composed of a red u-quark, a yellow u-quark and a green d-quark. One could imagine that the red u-quark and yellow u-quark come together first. Before they fly apart, a green d-quark comes at the right place to capture both u-quarks, so that they form a stable 'quantum spherical shell' with the d-quark at the center. One can determine the relativistic wave function of an u-quark based on the Dirac Hamiltonian with a confining linear potential, which is the solution of the gauge field with a fourth-order equation based on general Yang-Mills symmetry. It may also be possible for a collision between a red u-quark and a green d-quark to form a metastable system first, before capturing a yellow u-quark to form a stable proton. Other consecutive 2-body collisions might form neutrons and other baryons.

Consecutive 2-body collisions to form a 3-body system is the basic idea of the quantum shell model for baryons and anti-baryons. We demonstrate that, in such a quantum shell model for baryons, the green d-quark is acted upon by the quantum shell to create a new effective linear force, i.e., a harmonic oscillator potential, as discussed in Section 9-3 in Ch. 9. This idea was stimulated by a cosmological consideration of the linear Okubo force associated with baryonic charges acting on a supernova to produce the effects of 'dark energy'.[3] Thus, the C3Q model involves a harmonic oscillator potential V_{ho} with a constant of integration V_o,[1]

$$V_{ho}(r) \equiv Qr^2 + V_o, \tag{C1}$$

$$Q = \frac{16g_s^2\sqrt{Q'}R_\ell}{18L_s^2} \approx 20.28KR_\ell \times 10^8 \text{ MeV}^3, \quad R_\ell = \frac{\Gamma(\ell/2 + 7/2)}{\Gamma(\ell/2 + 2)},$$

$$R_0 = 3.3, \;\; R_1 = 4.5, \;\; R_2 = 5.8, \;\; R_3 = 7.2, \;\; R_4 = 8.7, \;\; R_5 = 10.3, \;\; R_6 = 12,$$

$$Q' = \frac{g_s^2 K^2}{8\pi L_s^2} = 20.2K^2 \times 10^4 \text{ MeV}^2,$$

$$1/fm = 197 \text{ MeV}, \quad L_s = 0.082 fm, \quad g_s^2/(4\pi) = 0.07, \quad c = \hbar = 1.$$

The constant V_o appears to be important for the spacing of the baryon energy eigenvalues of a relativistic Hamiltonian, as we shall see later.

A new feature of the C3Q model is the presence of an ℓ-dependent function in the harmonic oscillator potential, where ℓ denotes the angular momentum quantum number. The ℓ-dependent function R_ℓ is assumed to be an increasing function R_ℓ, which is suggested by the general pattern of the 29 energy states of the NJ^P baryons. The parameters V_o and K can be estimated from the particle data for baryons.[4] The quark coupling constant $g_s^2/4\pi$ and the basic length L_s are determined by charmonium data from the Cornell group.[5]

Thus the present C3Q model is based on two relativistic Hamiltonians H_u for the u-quarks and H_d for the d-quark, with respective masses $m_u = 2.16$ MeV and $m_d = 4.67$ MeV. The model involves two parameters V_o and K together with a 'correlation function' $R(\ell)$ to understand 29 highly relativistic eigenstates with energies ranging from 938 MeV up to 2700 MeV for the observed 3-quark N baryons.[4]

The solutions of the two relativistic u-quark wave functions suggest that the color charge density has the form,

$$|\Psi_u|^2 \propto exp(-a^2r^2)(a^2r^2)^{(K_0)}. \tag{C2}$$

(See equation (C13) below.)

In the quantum shell model for a proton (and other N baryons), the density of the charged shell roughly resembles the ground state of a helium atom. The model assumes that the two u-quarks in the shell do not contribute spin and orbital angular momentum to the proton. The total energy E of the 3-quark N baryon is assumed to be the sum of the relativistic energy E_d of the d-quark and the energies E_u of the two independent u-quarks in the shell, $E \approx E_d + 2E_u$.

C-2. Dirac Hamiltonian of the u-quark with a linear potential

Consider again our proton composed of red and yellow u-quarks, and a green d-quark. The C3Q model for 3-quark N baryons postulates that there are two separate interactions:

(i) Each of the two u-quarks moving around the d-quark with a linear potential $C(r)$ between them.

(ii) The d-quark moving in an effective harmonic oscillator potential, which is produced by the two u-quarks that are held together by a linear potential, acting on the d-quark. The physical properties of the relativistic d-quark dominate the properties of the N baryons.

Let us first consider the relativistic Hamiltonian[6] H_u of a u-quark and solve for the energy eigenvalues E_u with the help of the Sonine-Laguerre equation.[7] In the C3Q model, the Hamiltonian H_u with the linear potential $C(r) = Qr$ is postulated to be

$$H_u \approx \alpha_k p_k + \beta m + i\alpha_k e_k \beta C(r), \qquad (C3)$$

$$C(r) = \frac{g_s^2 K^2}{8\pi L_s^2} r \equiv Q'r, \quad Q' = 20.2K^2 \times 10^4 \text{ MeV}^2,$$

$$p_k = -i\partial/\partial x^k, \quad e_k = x_k/r, \quad k = 1, 2, 3,$$

$$\alpha_k = \begin{pmatrix} 0 & \sigma_k \\ \sigma_k & 0 \end{pmatrix}, \quad \beta = \begin{pmatrix} I & 0 \\ 0 & -I \end{pmatrix},$$

where α_k and β are the usual Dirac matrices.[6] The linear potential $C(r) = Q'r$ is approximated by the first term in (9.19) in Ch. 9. Since the coupling strength of the linear potential associated with baryons may not be the same as that of charmonium obtained by the Cornell group, we introduce a new parameter K in Q' to be determined by the baryon spectrum.

To find the energy eigenvalues of the quantum shell model, we write as usual the u-quark wave function q, which satisfies the equation

$$H_u q = Eq, \qquad (C4)$$

$$q = \begin{pmatrix} q_A \\ q_B \end{pmatrix} = \begin{pmatrix} g(r)r^{-1}Y_A \\ if(r)r^{-1}Y_B \end{pmatrix},$$

where $Y_A \equiv Y_{j\ell_A}^{j_3}$ and $Y_B \equiv Y_{j\ell_B}^{j_3}$ are r-independent normalized spin-angular functions (or spinor spherical harmonics).[6] The factor i in $if(r)r^{-1}Y_B$ is to make $f(r)$ and $g(r)$ real for bound state solutions. We have

$$(E - m)\frac{g(r)}{r}Y_A \approx (\sigma_k p_k - i\sigma_k e_k C)\frac{if(r)}{r}Y_B, \qquad (C5)$$

$$(E + m)\frac{if(r)}{r}Y_B \approx (\sigma_k p_k + i\sigma_k e_k C)\frac{g(r)}{r}Y_A.$$

As usual, we use the phase convention of Condon and Shortly,[6] so that we have $(\sigma_k r_k/r)Y_A = -Y_B$ and $(\sigma_k r_k/r)Y_B = -Y_A$, where Y_A and Y_B are the spinor spherical harmonics. Moreover, we also have the usual relations[6]

$$(\sigma_k p_k)q_B = i\frac{\sigma_k x_k}{r^2}\left(-ir\frac{\partial}{\partial r} + i\sigma_k L_k\right)\frac{f}{r}Y_B \qquad (C6)$$

$$= -\frac{d(f/r)}{dr}Y_A - \frac{(1 - \kappa)}{r^2}f(r)Y_A,$$

$$(\sigma_k p_k)q_A = i\frac{d(g/r)}{dr}Y_B + i\frac{(1 + \kappa)}{r^2}g(r)Y_B, \qquad (C7)$$

$$\kappa = (j + 1/2) = \ell > 0, \qquad \kappa = -(j + 1/2) = -(\ell + 1) < 0,$$

where L_k is the angular momentum operator. It follows from (C5)–(C7) that

$$\frac{df}{dr} - \frac{\kappa}{r}f + Cf \approx -(E - m)g, \qquad (C8)$$

$$\frac{dg}{dr} + \frac{\kappa}{r}g - Cg \approx (E + m)f, \qquad (C9)$$

in spherical coordinates. The conserved spin-orbital coupling quantum number κ is a non-zero integer which can be positive or negative. Roughly, the sign of κ determines whether the spin is antiparallel ($\kappa > 0$) or parallel ($\kappa < 0$) to the total angular momentum in the nonrelativistic limit.[6] The total angular momentum quantum number j and the angular momentum quantum number $\ell = \ell_A$ of the upper component q_A and the parity are determined by κ.

C-3. Basic equation for baryons and the energy eigenvalue equation

We obtain the r-dependent upper component $g = g(r)$ by eliminating the lower component $f = f(r)$ from equation (C8) by using (C9) and $C = Q'r$. This yields a basic equation for the d-quark in the quantum shell model,

$$\left[\frac{d^2}{dr^2} - \frac{\kappa^2 + \kappa}{r^2} + (2\kappa - 1)Q' - Q'^2r^2 + (E^2 - m^2)\right]g(r) \approx 0. \qquad (C10)$$

To find the energy eigenvalues of the u-quark, we define a new variable y, which is related to r^2 by

$$Q'r^2 = y. \tag{C11}$$

Based on the relations between κ and ℓ in (C7), we have $\kappa^2 + \kappa = \ell^2 + \ell$ for both $\kappa = (j + 1/2) = \ell > 0$, and $\kappa = -(j + 1/2) = -(\ell + 1) < 0$. Thus, equation (C10) can be written as

$$\left[y\frac{d^2}{dy^2} + \frac{1}{2}\frac{d}{dy} - \frac{\ell^2 + \ell}{4y} - \frac{y}{4} + \frac{(2\kappa - 1)}{4} + \frac{E^2 - m^2}{4Q'} \right] g(y) \approx 0. \tag{C12}$$

We look for a solution to this equation of the form[6]

$$g(r) = e^{-y/2} y^{[(\ell+1)/2]} G(y), \tag{C13}$$

which is consistent with the asymptotic property $g(y) \to 0$ as $y \to \infty$.

We obtain the following Sonine-Laguerre differential equation[7] for $G(y)$,

$$\left[y\frac{d^2}{dy^2} + \left(\ell + \frac{3}{2} - y \right)\frac{d}{dy} - \frac{\ell}{2} - \frac{3}{4} + \frac{(2\kappa - 1)}{4} + \frac{E^2 - m^2}{4Q'} \right] G(y) \approx 0. \tag{C14}$$

The solutions to the equation (C14) are the Sonine-Laguerre polynomials of degree n. Note that n comes from the Sonine-Laguerre equation $y d^2G/dy^2 + (A - y)dG/dy + nG = 0$, and n is an integer greater than or equal to zero. Thus, we have

$$n = -\left(\frac{\ell}{2} + \frac{3}{4} - \frac{2\kappa - 1}{4} - \frac{E^2 - m^2}{4Q'} \right), \quad E = E_u, \quad m = m_u, \tag{C15}$$

where $n = 0, 1, 2, ...$ The confining 3-quark (C3Q) model gives the energy eigenvalues of one u-quark $E_u = E$ in terms of the principal quantum number n, orbital angular momentum quantum number ℓ, and a 'spin-orbit' coupling quantum number κ,

$$E_u = \left[4Q'\left(n + \frac{\ell - \kappa}{2} + 1 \right) + m_u^2 \right]^{0.5}. \tag{C16}$$

C-4. Relativistic Hamiltonian and energy eigenvalues of the d-quark in a quantum shell

Apart from the energies of two u-quarks, the C3Q model assumes that the relativistic motion of the d-quark also contributes to the proton mass. The model postulates the following Hamiltonian for the d-quark, H_d, moving in the quantum shell of the two u-quarks,

$$H_d \approx \alpha_k p_k + \beta m_d + \frac{(1 + \beta)}{2} V_{ho}, \quad V_{ho} = Qr^2 + V_o. \tag{C17}$$

We first derive the effective harmonic oscillator potential V_{ho} produced by the two u-quarks in the quantum shell of a proton. We follow the steps in Section 9-3 in Ch. 9, except we replace $\Psi = (1/a_q^{3/2}/2)exp(-R'/a_q)$ in equation (9.21) by $\Psi' \propto [exp(-a^2r^2/2)](a^2r^2)^{(\ell+k)/2}$, as suggested by the wave function (C13), and we also introduce two parameters K and V_o to be determined by the baryon spectrum. To obtain the energy eigenvalue E_d in (C17) of the d-quark, we use H_d in (C17) and follow the steps from (C3) to (C16). Instead of equations (C8), (C9) and (C10), we have the following equations for the d-quark,

$$\frac{df}{dr} - \frac{\kappa}{r}f \approx -(E - m - V_{ho})g, \tag{C18}$$

$$\frac{dg}{dr} + \frac{\kappa}{r}g \approx (E + m)f. \tag{C19}$$

$$\left[\frac{d^2}{dr^2} - \frac{\kappa^2 + \kappa}{r^2} - (E + m)V_{ho} + (E^2 - m^2)\right]g(r) \approx 0. \tag{C20}$$

Since $V_{ho} = Qr^2 + V_o$, it is convenient to define a new variable $y = \sqrt{[(E + m)Q]}\, r^2 \equiv a^2r^2$. Equation (C20) can then be written as

$$\left[4y\frac{d^2}{dy^2} + 2\frac{d}{dy} - \frac{\ell^2 + \ell}{y} - y + \frac{E^2 - m^2 - (E + m)V_o}{a^2}\right]g(y) \approx 0. \tag{C21}$$

We look for a solution of the form, $g(r) = e^{-y/2}y^{[(\ell+1)/2]}G'(y)$ for (C21). We obtain the Sonine-Laguerre equation[7] for the d-quark in the harmonic oscillator (HO) potential V_{ho},

$$\left[y\frac{d^2}{dy^2} + \left(\ell + \frac{3}{2} - y\right)\frac{d}{dy} - \frac{\ell}{2} - \frac{3}{4} + \frac{E^2 - m^2 - (E + m)V_o}{4a^2}\right]G'(y) \approx 0, \tag{C22}$$

$$a^2 = \sqrt{[(E + m)Q]}.$$

Following steps (C12)–(C16), we obtain the relativistic equation for the energy eigenvalue $E = E_d, m = m_d$ of the d-quark,

$$E_d^2 - m_d^2 = 4\sqrt{(E_d + m_d)Q}\left(n + \frac{\ell}{2} + 0.75 + b\right), \tag{C23}$$

$$Q = \frac{16g_s^2\sqrt{Q'}R_\ell}{18L_s^2} \approx 20.28KR_\ell \times 10^8 \text{ MeV}^3, \quad b = \frac{(E_d + m_d)V_o}{4a^2},$$

where $1/L_s = 24.02 \times 10^2$ MeV, and V_o is expressed in terms of a dimensionless parameter b for convenience.

C-5.   The C3Q model and the energy spectrum of N baryons

In 2020, there were 29 N baryon energy states listed in the particle data. The existence of 22 of them is deemed very likely or certain, and in general, their properties are fairly well-explored. Much less is known about the other 7 states, and their existence is much less certain, according to the particle data group P. A. Zyla et al.[4]

Based on (C16) and (C23), the C3Q model gives the energy eigenstate $E_n(N)$ of an N baryon as the sum of the d-quark energy E_d and the two u-quark energy E_{2u},

$$E_n(N) \approx E_d + E_{2u}, \qquad (C24)$$

where

$$E_{2u} = 2\sqrt{\left[4Q'\left(n + \frac{\ell - \kappa}{2} + 1\right) + m_u^2\right]}, \quad m_u = 2.16 \text{ MeV},$$

and E_d is the positive energy solution of the equation (C23), or

$$E_d = \sqrt{4\sqrt{(E_d + m_d)Q}\left(n + \frac{\ell}{2} + 0.75 + b\right) + m_d^2}, \quad m_d = 4.67 \text{ MeV}.$$

Let us consider the energies of the two lowest N baryon states, i.e., the proton and neutron, and their energy difference. The proton mass $m(p)$ is given by (C24) with n = 0, $\kappa = -(j + 1/2) = -1$ or $\kappa = -(\ell + 1) = -1$ or $\ell = 0$ (i.e., positive parity). Using WolframAlpha, we obtain

$$m(p^+) = E_d + E_{2u} \approx 932.5 + 6.2 \approx 938.7 \text{ MeV}, \qquad \ell = 0, \ b = 1, \quad (C25)$$

$$Q \approx 20.28 K R_\ell \times 10^8 \text{ MeV}^3, \qquad K = 2.46 \times 10^{-3}, \qquad R_\ell = \frac{\Gamma(\ell/2 + 7/2)}{\Gamma(\ell/2 + 2)},$$

where $n = 0$ and $R_\ell = 3.3$ for $\ell = 0$. Based on E_{2u} and E_d in (C24), we estimate the values of Q and K associated with the effective harmonic oscillator potential of the d-quark.

For a neutron (udd), the C3Q model implies red and yellow d-quarks forming a color charge quantum shell that generates the harmonic oscillator potential $V_{HO} = Qr^2 + V_o$. A green u-quark sits near the center of this potential. Note that the relativistic mass of the u-quark in the harmonic oscillator potential is about 931 MeV, which is much larger than the d-quark mass 4.67 MeV in the shell of a neutron. To be specific, instead of E_{2u} and E_d in (C24) for a proton, we use the following energy eigenvalue $E(n^0)$ for a neutron,

$$E(n^0) \approx E_u + E_{2d}, \qquad (C26)$$

where

$$E_{2d} = 2\sqrt{\left[4Q'\left(n + \frac{\ell - \kappa}{2} + 1\right) + m_d^2\right]}, \quad m_d = 4.67 \text{ MeV}$$

and E_u is the positive energy solution of

$$E_u = \sqrt{4\sqrt{(E_u + m_u)Q}\left(n + \frac{\ell}{2} + 0.75 + b\right) + m_u^2}, \quad m_u = 2.16 \text{ MeV}.$$

The energy eigenvalue equations (C26) with $n = 0, \ell = 0, \kappa = -1$ and $b = 1$ give the neutron mass $m(n^0) = E(n^0)$,

$$m(n^0) \approx E_u + E_{2d} \approx 931.6 + 10.8 = 942.4 \text{ MeV}, \quad \text{(C27)}$$

$$m(n^0) - m(p^+) \approx 3.7 \text{ MeV},$$

$$Q \approx 20.28 K R_\ell \times 10^8 \text{ MeV}^3, \quad K = 2.46 \times 10^{-3}, \quad R_0 = 3.3,$$

where the corresponding experimental values are 939.5 MeV and 1.3 MeV, respectively. These are much better results than previous estimations based on dispersion relations.[a] The neutron and the proton are isospin doublets and naturally, have the same color potentials with the same parameters K and b.

For $N^+(uud)$ baryon states with positive charge and even parity, i.e., $N1/2^+$, the C3Q model gives the following predictions for $E_n \approx E_d(n) + E_{2u}(n)$ with the corresponding experimental values[4]

$$model: \quad n = 0, \quad E_0(1/2^+) \approx 938.6 \text{ MeV}, \quad data: (938.3)1/2^+. \quad \text{(C28)}$$

$$n = 1, \quad E_1(1/2^+) \approx 1259 \text{ MeV}, \qquad N(1440)1/2^+,$$

$$n = 2, \quad E_2(1/2^+) \approx 1548 \text{ MeV}, \qquad N(1710)1/2^+,$$

$$n = 3, \quad E_3(1/2^+) \approx 1812 \text{ MeV}, \qquad N(1880)1/2^+,$$

$$n = 4, \quad E_4(1/2^+) \approx 2058 \text{ MeV}, \qquad N(2100)1/2^+,$$

$$n = 5, \quad E_5(1/2^+) \approx 2291 \text{ MeV}, \qquad N(2300)1/2^+,$$

$$K = 2.46 \times 10^{-3}, \quad b = 1, \quad \ell = 0, \quad R_0 = 3.3.$$

The notation $1/2^+$ in the experimental data denotes J^P, where $J = 1/2$ is the total angular momentum quantum number and P is parity.[6]

---

[a]See discussions after equation (9.34) in Ch. 9.

For N^0 baryons, which consist of 'udd' rather than 'uud', we use the same parameters as those in (C27) for a neutron, i.e., Q, K, and R_0 in (C27). The eigenvalue equations in (C26) give the following predictions for the $N^0(udd)$ baryons (with + parity and energy $E'_n(1/2^+) \approx E_u(n) + E_{2d}(n)$),

$$\begin{array}{llll}
model: n = 0, & E'_0(1/2^+) = 942.4 \text{ MeV}, & data: & 939.5 \text{ MeV}, & \text{(C29)} \\
n = 1, & E'_1(1/2^+) \approx 1270 \text{ MeV}, & & N(1440)1/2^+, \\
n = 2, & E'_2(1/2^+) \approx 1560 \text{ MeV}, & & N(1710)1/2^+, \\
n = 3, & E'_3(1/2^+) \approx 1825 \text{ MeV}, & & N(1880)1/2^+, \\
n = 4, & E'_4(1/2^+) \approx 2072 \text{ MeV}, & & N(2100)1/2^+, \\
n = 5, & E'_5(1/2^+) \approx 2305 \text{ MeV}, & & N(2300)1/2^+, \\
\end{array}$$
$$K = 2.46 \times 10^{-3}, \quad b = 1, \quad \kappa = -1, \quad \ell = 0, \quad R_0 = 3.3.$$

As one can see, these energy eigenvalues are close to the energies of the corresponding N^+ baryons in (C28), just as for the proton and the neutron.

Predictions for other N^+ ('uud') baryon states $E_n(N3/2^+)$ with $J^P = 3/2^+$ are given by the C3Q model as follows:

$$\begin{array}{lll}
E_0(N3/2^+) \approx 1526 \text{ MeV}, & N(1720)3/2^+, & \text{(C30)} \\
E_1(N3/2^+) \approx 1876 \text{ MeV}, & N(1900)3/2^+, \\
E_2(N3/2^+) \approx 2191 \text{ MeV}, & N(2040)3/2^+, \\
\end{array}$$
$$K = 2.46 \times 10^{-3}, \kappa = 2 \quad b = 1, \quad \ell = 2, \quad R_2 = 5.8,$$

where the parameters K and R_ℓ are the same as those used in (C25) for protons. We note that the evidence of the existence of the state $N(2040)3/2^+$ in (C30) is poor. In this connection, the C3Q model suggests a new state $E_2(3/2^+) \approx 2191$ MeV to replace $N(2040)3/2^+$, which has poor evidence.[4]

To further test the C3Q model, let us consider the energy spectrum of $\Sigma^+(uus)$, which involves two u-quarks (say, red and yellow) and one strange quark s (green) with mass $m_s = 93$ MeV. To obtain a rough numerical estimate, we use the same parameters as those for the N baryons. In this case, a different value for the parameter K as shown below is determined by the ground state with mass 1189 MeV, i.e., $\Sigma(1189)1/2^+$. Using (C24) with E_d and m_d replaced respectively by E_s and $m_s = 93$ MeV, we obtain the following approximate results,

$$\begin{array}{lll}
E_0(\Sigma 1/2^+) \approx 1195 \text{ MeV}, & \Sigma(1189)1/2^+, & \text{(C31)} \\
E_1(\Sigma 1/2^+) \approx 1602 \text{ MeV}, & \Sigma(1660)1/2^+, \\
E_2(\Sigma 1/2^+) \approx 1960 \text{ MeV}, & \Sigma(1880)1/2^+, \\
\end{array}$$
$$\ell = 0, \quad K = 4.64 \times 10^{-3}, \quad b = 1, \quad R_0 = 3.3.$$

Note that the parameter $K = 4.64 \times 10^{-3}$ for $E_n(\Sigma 1/2^+)$(with u u s) differs from that in $E_n(N)$ for N baryons (with u u d) because s and d are different quarks.

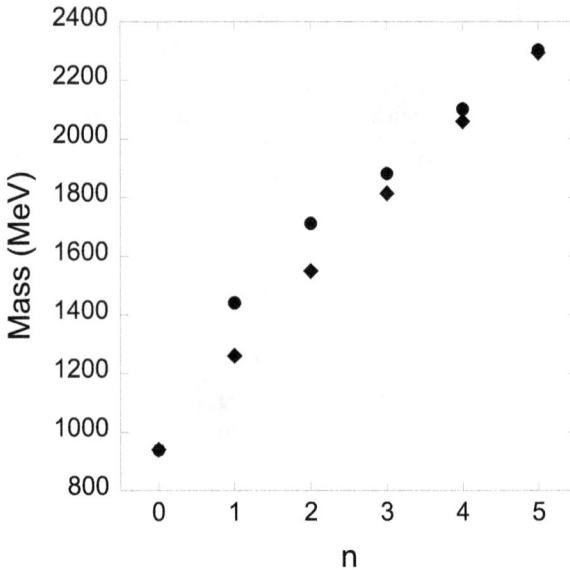

Fig. C.1. The energy spectrum of N baryons $E_n(1/2^+)$. Predictions of the model are shown by diamonds and measured masses are indicated by circles. The experimentally measured $n = 0$ baryon mass is used to set the values of the model parameters K and b (or V_o) in (C23). (Thus the exact agreement for that mass).

C-6. N baryon states with negative parity and large total angular momenta

The established baryon states with negative parity are[4]

$$N(1535)1/2^-, N(1650)1/2^-, N(1895)1/2^- \qquad (C32)$$

$$N(1520)3/2^-, N(1700)3/2^-, N(1875)3/2^-, N(2120)3/2^-.$$

Their existences are certain and their properties are at least fairly well explored.[4]

We can use the eigenvalue equation (C24) of the C3Q model to understand them. We have the results

$$E_0(N1/2^-) \approx 1221 \text{ MeV}, \qquad N(1535)1/2^-, \qquad (C33)$$

$$E_1(N1/2^-) \approx 1560 \text{ MeV}, \qquad N(1650)1/2^-,$$

$$E_2(N1/2^-) \approx 1865 \text{ MeV}, \qquad N(1895)1/2^-,$$

where we have used the following values for the parameters,

$$\ell = 1 = \kappa, \quad K = 2.46 \times 10^{-3}, \quad b = 1, \quad R_1 = 4.5. \qquad \text{(C34)}$$

Similarly, for baryon states with $N(1520)3/2^-$ etc. in (C32), we obtain the approximate results,

$$E_0(N3/2^-) \approx 1222 \text{ MeV}, \qquad N(1520)3/2^-, \qquad \text{(C35)}$$

$$E_1(N3/2^-) \approx 1561 \text{ MeV}, \qquad N(1700)3/2^-,$$

$$E_2(N3/2^-) \approx 1866 \text{ MeV}, \qquad N(1875)3/2^-,$$

$$E_3(N3/2^-) \approx 2148 \text{ MeV}, \qquad N(2120)3/2^-,$$

where we have used the following values for parameters in (C26),

$$\ell = 1, \quad \kappa = -2, \quad K = 2.46 \times 10^{-3}, \quad b = 1, \quad R_1 = 4.5.$$

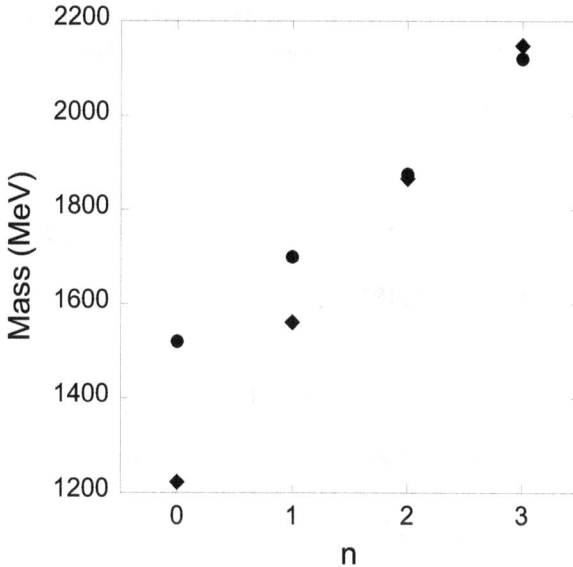

Fig. C.2. The energy spectrum of N baryons $E_n 3/2^-$ given in (C35). Predictions of the model are shown by diamonds and measured masses are indicated by circles. The wide gap between the model and the data for the baryon mass E_n for $n = 0$ shows the approximate nature of the quantum shell model.

Let us consider the energy spectrum of $\Sigma^+(uus)$ with negative parity,[b] which involves two u-quarks (say, red and yellow) and one strange quark s (green) with mass $m_s = 93$ MeV. We expect the value of K will be different from the previous case for the proton(uud). Using the eigenvalue equation (C24) with E_d and m_d replaced respectively by E_s and $m_s = 93$ MeV, we obtain

$$E_0(\Sigma 1/2^-) \approx 1541 \text{ MeV}, \qquad \Sigma(1620)1/2^-, \qquad (\text{C36})$$

$$E_1(\Sigma 1/2^-) \approx 1960 \text{ MeV}, \qquad \Sigma(1900)1/2^-,$$

$$E_2(\Sigma 1/2^-) \approx 2337 \text{ MeV}, \qquad \Sigma(2160)1/2^-,$$

$$K = 4.64 \times 10^{-3}, \quad \ell = 1, \quad b = 1, \quad R_1 = 4.5,$$

where the value of the parameter K is the same as that in (C31). Note that R_ℓ takes a different value from that in (C31) due to different parity (i.e., different ℓ).

To have a complete picture of N baryons from $J^P = 1/2^+$ to $J^P = 11/2^-$ and $J^P = 13/2^+$, let us display their results based on the energy eigenvalue equation (C24) and the corresponding experimental data:

$$N5/2^+, \quad \ell = 2, \quad K = 2.46 \times 10^{-3}, \quad R_2 = 5.8, \qquad (\text{C37})$$

$$E_0(N5/2^+) \approx 1520 \text{ MeV}, \qquad N(1680)5/2^+,$$

$$E_1(N5/2^+) \approx 1868 \text{ MeV}, \qquad N(1860)5/2^+,$$

$$E_2(N5/2^+) \approx 2187 \text{ MeV}, \qquad N(2000)5/2^+.$$

$$N5/2^-, \quad \ell = 3, \quad K = 2.46 \times 10^{-3}, \quad R_3 = 7.2, \qquad (\text{C38})$$

$$E_0(N5/2^-) \approx 1825 \text{ MeV}, \qquad N(1675)5/2^-,$$

$$E_1(N5/2^-) \approx 2183 \text{ MeV}, \qquad N(2060)5/2^-,$$

$$E_2(N5/2^-) \approx 2513 \text{ MeV}, \qquad N(2570)5/2^-.$$

$$N7/2^+, \quad \ell = 4, \quad K = 2.46 \times 10^{-3}, \quad R_4 = 8.7, \qquad (\text{C39})$$

---

[b]This spectrum has a different parity from that in (C31).

$$E_0(N7/2^+) \approx 2139 \text{ MeV}, \qquad N(1990)7/2^+.$$

$$N7/2^-, \quad \ell = 3, \quad K = 2.46 \times 10^{-3}, \quad R_3 = 7.2, \qquad \text{(C40)}$$

$$E_0(N7/2^-) \approx 1825 \text{ MeV}, \qquad N(2190)7/2^-.$$

$$N9/2^+, \quad \ell = 4, \quad K = 2.46 \times 10^{-3}, \quad R_4 = 8.7, \qquad \text{(C41)}$$

$$E_0(N9/2^+) \approx 2139 \text{ MeV}, \qquad N(2220)9/2^+.$$

$$N9/2^-, \quad \ell = 5, \quad K = 2.46 \times 10^{-3}, \quad R_5 = 10.3, \qquad \text{(C42)}$$

$$E_0(N9/2^-) \approx 2459 \text{ MeV}, \qquad N(2250)9/2^-.$$

$$N11/2^-, \quad \ell = 5, \quad K = 2.46 \times 10^{-3}, \quad R_5 = 10.3, \qquad \text{(C43)}$$

$$E_0(N11/2^-) \approx 2459 \text{ MeV}, \qquad N(2600)11/2^-.$$

$$N13/2^+, \quad \ell = 6, \quad K = 2.46 \times 10^{-3}, \quad R_6 = 12, \qquad \text{(C44)}$$

$$E_0(N13/2^+) \approx 2787 \text{ MeV}, \qquad N(2700)13/2^+,$$

where b has the same value, $b = 1$, in (C37)–(C44).

C-7. Quantum shell model applied to 3-antiquark antibaryons

To apply the C3Q model to antimatter, all quarks in the previous discussions are replaced by their corresponding antiquarks. To be specific, the (uud) quarks in the Hamiltonians H_u and H_d in (C3) and (C17) are replaced by the antiquarks $(\bar{u}\,\bar{u}\,\bar{d})$, so that we have the Hamiltonians $H_{\bar{u}}$ and $H_{\bar{d}}$,

$$H_{\bar{u}} \approx \alpha_k p_k + \beta m_{\bar{u}} + i\alpha_k e_k \beta C(r), \quad C(r) = \frac{g_s^2 K^2}{8\pi L_s^2} r = Q'r, \qquad \text{(C45)}$$

$$H_{\bar{d}} \approx \alpha_k p_k + \beta m_{\bar{d}} + \frac{(1+\beta)}{2} V_{ho}, \quad V_{ho} = Qr^2 + V_o,$$

$$Q = 20.28 K R_\ell \times 10^8 \text{ MeV}^3, \quad R_\ell = \frac{\Gamma(\ell/2 + 7/2)}{\Gamma(\ell/2 + 2)}.$$

Following the same steps of previous calculations, we have the resultant energy eigenvalues,

$$E_{\overline{u}} = \left[4Q'\left(n + \frac{\ell - \kappa}{2} + 1 \right) + m_{\overline{u}}^2 \right]^{0.5}, \tag{C46}$$

$$E_{\overline{d}}^2 - m_{\overline{d}}^2 = 4\sqrt{(E_{\overline{d}} + m_{\overline{d}})Q}\left(n + \frac{\ell}{2} + 0.75 + b \right),$$

$$Q = \frac{8 g_s^2 \sqrt{Q'} R_\ell}{9 L_s^2} \approx 20.28 K R_\ell \times 10^8 \text{ MeV}^3, \quad b = \frac{(E_{\overline{d}} + m_{\overline{d}})V_o}{4a^2}.$$

Thus, all the energy eigenvalues for baryons of the model in (C25)–(C44) can be used for understanding antibaryons, and by extension, the evolution of the anti-matter half-universe in the Big Jets model.

C-8. Confining 2-quark model for the K meson spectrum

Since the C3Q model seems to give reasonable predictions for the energy states (masses) of 3-quark baryons, it is natural to see whether one can understand the $q\overline{q}'$ spectrum of mesons based on a linear potential similar to $C(r)$ in (C3). It turns out that the $q\overline{q}'$ energy eigenstates of mesons seem to be more 'complicated' to understand than that of baryons because the meson spectrum appears to be more irregular than the baryon spectrum based on a Dirac Hamiltonian.

Nevertheless, it is interesting to investigate the non-trivial quantum 2-body problems of mesons. Consider K mesons for example. The spectra of $K^+(u\overline{s})$ and $K^0(d\overline{s})$ seem relatively simple because they are all J^P states with the same isospin, in contrast to the light unflavored mesons, e.g., π mesons, etc.

If one assumes the u-quark u and the anti-strange quark $\overline{s}$ in a K^+ meson satisfy a Dirac Hamiltonian with only a linear confining potential, the resultant energy eigenstates are very much different from the experimental data. Thus it appears unlikely that one can understand the spectrum of K^+ mesons based solely on a linear confining potential.

However, the experimental data suggests that a better dynamical picture for the K meson system can be obtained by the following assumptions:

(i) The light u-quark in K mesons produces a quantum shell with an effective harmonic oscillator potential $V_{ho} = Qr^2 + V_o$, whereas the heavy anti-s-quark, $\bar{s}$ is located near the center of the potential and provides the dominant contribution to the energy of the K^+. The parity P of a K meson state in the model is defined to be related to the orbital quantum number ℓ of the K meson state, i.e. $P = (-1)^{\ell+1}$.[c]

(ii) The u-quark moves in a linear potential $V = Q'r$ and its energy contributes only a few percents to the energy (or mass) of a K meson.

Interestingly, such a picture resembles the quantum shell of the C3Q model for baryons. In this sense, we have a very roughly unified picture for the dynamics of N baryons and (heavy) K mesons.

The principal quantum number n of a K meson state is usually defined to have the values[4] 1, 2, 3.... However, in the present model, it is natural to define $n = 0, 1, 2, ...$, because n in (C15) came from the Sonine-Laguerre equation $yd^2G/dy^2 + (A - y)dG/dy + nG = 0$, where n is an integer greater than or equal to zero. It appears that the quantum numbers n and ℓ are model dependent. (See also footnote in Section 8.) Here, the K mesons are states of $q\bar{q}'$ systems, in which the parity is $(-1)^{\ell+1}$, where ℓ is the angular momentum of the state.[4]

The $\bar{s}$ quark Hamiltonian $H_{\bar{s}}$ is assumed to be formally the same as that in (C17) and hence, its energy eigenvalues $E_{\bar{s}}$ are formally the same as those in (C23),

$$H_{\bar{s}} \approx \alpha_k p_k + \beta m_{\bar{s}} + \frac{(1+\beta)}{2} V'_{ho}, \qquad V'_{ho} = Q_m r^2 + V'_o, \qquad (C47)$$

$$E_{\bar{s}}^2 - m_{\bar{s}}^2 = 4\sqrt{(E_{\bar{s}} + m_{\bar{s}})Q_m} \left(n + \frac{\ell}{2} + 0.75 + b'\right), \qquad (C48)$$

$$Q_m = 20.28K'R_\ell \times 10^8 \text{ MeV}^3, \qquad R_\ell = \frac{\Gamma(\ell/2 + 7/2)}{\Gamma(\ell/2 + 2)},$$

$$b' = \frac{(E_{\bar{s}} + m_{\bar{s}})V'_o}{4a'^2}, \qquad a'^2 = \sqrt{(E_{\bar{s}} + m_{\bar{s}})Q_m},$$

where V'_o is expressed in terms of a dimensionless parameter b' for convenience.

The u-quark Hamiltonian H_u is assumed to be the same as that in (C3),

$$H_u \approx \alpha_k p_k + \beta m + i\alpha_k e_k \beta C(r). \qquad (C49)$$

---

[c]Usually, if the orbital angular momentum of the $q\bar{q}'$ state is ℓ, then the parity[4] P is $(-1)^{\ell+1}$. The K mesons with larger masses appear to be more in line with the energy eigenstates of the Hamiltonian $H_{\bar{s}}$ of the $\bar{s}$ quark, as shown in (C60)–(C66), which have the same value for b'. This property is the same as that for the 3-quark baryons as shown in (C25)–(C44), they all have the same values $b = 1$.

Its energy eigenvalues E_u can be obtained similarly using the procedure in Sections 2 and 3,

$$E_u = \left[4Q'\left(n + \frac{\ell - \kappa}{2} + 1\right) + m_u^2\right]^{0.5}, \quad Q' = 20.2K'^2 \times 10^4 \text{ MeV}^2. \quad (C50)$$

The total energy of the K^+, which involves u and $\bar{s}$, is assumed to be given by the sum,

$$E(K^+) \approx E_{\bar{s}}(K^+) + E_u(K^+). \quad (C51)$$

The $J^P = 0^-$ data, $K^+(490), K^+(1460), K^+(1830)$, can be estimated by using the eigenvalues $E_{\bar{s}}$ and E_u in (C48) and (C50). We obtain

$$model: \quad E_0 \approx 502 \text{ MeV}, \qquad data: \quad K^+(493.7)0^-, \qquad (C52)$$

$$E_1 \approx 1048 \text{ MeV}, \qquad\qquad K^+(1460)0^-,$$

$$E_2 \approx 1468 \text{ MeV}, \qquad\qquad K^+(1830)0^-,$$

where we have used the following parameters,

$$Q_m = 20.28K'R_0, \quad K' = K = 0.0046, \quad \ell = 0, \quad R_0 = 3.3, \quad (C53)$$

$$b' = -0.3, \quad m_u = 2.16 \text{ MeV}, \quad m_s = 93 \text{ MeV}.$$

We have the relation $K' = K = 0.0046$ because both $K^+(u\bar{s})$ mesons in (C52) and $\Sigma^+(uus)$ in (C36) involve the strange quark.

Similarly, using (C48) and (C50) with the u-quark replaced by a d-quark, we have the following approximate results for $K^0(d\bar{s})$,

$$model: \quad E_0 \approx 502 \text{ MeV}, \qquad data: \quad K^0(497.6)0^-, \qquad (C54)$$

$$E_1 \approx 1046 \text{ MeV}, \qquad\qquad K^0(1460)0^-,$$

$$E_2 \approx 1468 \text{ MeV}, \qquad\qquad K^0(1830)0^-,$$

where we have used $m_d = 4.67$ MeV and the same parameters as those in (C53). From (C50)–(C54), we obtain the approximate mass difference of K^0 and K^+,

$$m(K^0) - m(K^+) \approx 1.6 \text{ MeV}, \quad (C55)$$

where the experimental value is about 4.1 MeV. The mass difference in (C55) is mainly determined by the energy eigenvalue (C50) for a u-quark and that for a d-quark (which is the same as (C50) with u replaced by d).[d]

---

[d]The quantum number κ in (C50) is neglected because its relation to j and ℓ of a K meson is not clear in the simplified model. In contrast, the relation is clear for the N baryons, as shown in (C7).

For a big picture of the non-trivial pattern of the K meson states, we display the K meson spectra from $K^*(700)0^+$ to $K(3000)5^+$. For $K0^+$ energy eigenstates, we have the following approximate results based on (C48),

$$K0^+, \quad \ell = 1, \quad K' = 0.0046, \quad R_1 = 4.5, \quad b' = -0.7; \qquad \text{(C56)}$$

$$model: \quad E_0(K0^+) \approx 624 \text{ MeV}, \qquad data: \quad K^*(700)0^+,$$

$$E_1(K0^+) \approx 1207 \text{ MeV}, \qquad\qquad K^*(1430)0^+,$$

$$E_2(K0^+) \approx 1667 \text{ MeV}, \qquad\qquad K^0(1630)?^?,$$

$$E_3(K0^+) \approx 2070 \text{ MeV}, \qquad\qquad K^0(1950)0^+,$$

$$K1^-, \quad \ell = 0, \quad K' = 0.0046, \quad R_0 = 3.3, \quad b' = +0.3; \qquad \text{(C57)}$$

$$E_0 \approx 851 \text{ MeV}, \qquad\qquad K^*(892)1^-,$$

$$E_1 \approx 1308 \text{ MeV}, \qquad\qquad K^*(1410)1^-,$$

$$E_2 \approx 1694 \text{ MeV}, \qquad\qquad K^*(1680)1^-,$$

$$K1^+, \quad \ell = 1, \quad K' = 0.0046, \quad R_1 = 4.5, \quad b' = -0.3; \qquad \text{(C58)}$$

$$E_0 \approx 881 \text{ MeV}, \qquad\qquad K_1(1270)1^+,$$

$$E_1 \approx 1400 \text{ MeV}, \qquad\qquad K_1(1400)1^+,$$

$$E_2 \approx 1834 \text{ MeV}, \qquad\qquad K_1(1650)1^+,$$

$$K2^-, \quad \ell = 2, \quad K' = 0.0046, \quad R_2 = 5.8, \quad b' = -0.5 \qquad \text{(C59)}$$

$$E_0 \approx 1140 \text{ MeV}, \qquad\qquad K_2(1580)2^-,$$

$$E_1 \approx 1669 \text{ MeV}, \qquad\qquad K_2(1770)2^-,$$

$$E_2 \approx 2123 \text{ MeV}, \qquad\qquad K_2(1820)2^-,$$

$$E_3 \approx 2572 \text{ MeV}, \qquad\qquad K_2(2250)2^-,$$

$$K2^+, \quad \ell = 1, \quad K' = 0.0046, \quad R_1 = 4.5, \quad b' = +0.7; \qquad \text{(C60)}$$

$$E_0 \approx 1400 \text{ MeV}, \qquad K_2^*(1430)2^+,$$

$$E_1 \approx 1834 \text{ MeV}, \qquad K_2^*(1980)2^+,$$

$$K3^-, \quad \ell = 2, \quad K' = 0.0046, \quad R_2 = 5.8, \quad b' = +0.7; \qquad \text{(C61)}$$

$$E_0 \approx 1793 \text{ MeV}, \qquad K_3^*(1780)3^-,$$

$$K3^+, \quad \ell = 3, \quad K' = 0.0046, \quad R_3 = 7.2, \quad b' = +0.7; \qquad \text{(C62)}$$

$$E_0 \approx 2139 \text{ MeV}, \qquad K_3(2320)3^+,$$

$$K4^-, \quad \ell = 4, \quad K' = 0.0046, \quad R_4 = 8.7, \quad b' = +0.7; \qquad \text{(C63)}$$

$$E_0 \approx 2523 \text{ MeV}, \qquad K_4(2500)4^-,$$

$$K4^+, \quad \ell = 3, \quad K' = 0.0046, \quad R_3 = 7.2, \quad b' = +0.7; \qquad \text{(C64)}$$

$$E_0 \approx 2139 \text{ MeV}, \qquad K_4^*(2045)4^+,$$

$$K5^-, \quad \ell = 4, \quad K' = 0.0046, \quad R_4 = 8.7, \quad b' = +0.7; \qquad \text{(C65)}$$

$$E_0 \approx 2523 \text{ MeV}, \qquad K_4^*(2380)5^-,$$

$$K5^+, \quad \ell = 5, \quad K' = 0.0046, \quad R_5 = 10.3, \quad b' = +0.7; \qquad \text{(C66)}$$

$$E_0 \approx 2916 \text{ MeV}, \qquad K(3100)?^?.$$

The simplified model suggests that the K meson[4] $K(3100)?^?$ in (C66) has the isospin quantum numbers $1/2$, and $J^P = 5^+$.

C-9. Discussion and conclusion

In the C3Q model, a new quark Hamiltonian leads to the Sonine-Laguerre equation,[7] which gives analytically the energy eigenvalues of the N baryons. Based solely on two parameters, the coupling strength K (i.e., Q) and a constant of integration V_o, together with a 'correlation' function $R_\ell = \Gamma(\ell/2 + 7/2)/\Gamma(\ell/2 + 2)$, the highly simplified C3Q model achieves agreements within 0.1% to 16% for the 29 N baryon energy eigenvalues. For the $\Sigma^+(uus)$ spectrum in equation (C30), a single new value of the coupling constant involving the parameter $K = 4.64 \times 10^{-3}$ enables the fitting of three Σ^+ baryon masses in the range of 1190 MeV to 1880 MeV with deviations of ≈ 0.6–8%. These results suggest that there may be some truth to the new quark Hamiltonians H_d and H_u, however simplified they may be.

The totality of roughly 170 baryons of all types has a more complicated pattern of energy eigenstates.[4] For example, Λ baryons consists of 3 different quarks (u d s), whose quantum shell structure appears to be more complicated than that of N baryons with 3 quarks (u u d). The mass problem of baryons is a quantum 3-body problem, and predicting the mass of elementary particles is perhaps an ultimate challenge to physicists' imaginations.

To see the non-trivial pattern of the meson states, we also tried to use the quantum shell model to fit the charmonium spectra,[5] from $J/\Psi(3096)1^-$ to $\Psi(4660)1^-$ based on the relativistic eigenvalue equation (C48) with different values for parameters. The result appears to be similar to (or worse than) that of the K meson spectrum in (C56)–(C66).

Interestingly, the results in section C-8 for the K meson spectrum reveal a certain similarity between the dynamics of N baryons and that of K mesons in the quantum shell model. Both have quark Hamiltonians with effective harmonic oscillator and linear potentials. It is puzzling that the parameter b' (or V_o) takes the value $b' = +0.7$ for the K meson states from $K_2^*(1430)2^+$ in (C60) to $K5^+$ or K(3100) in (C66), but takes different values in (C56)–(C59) for K mesons with smaller masses. This property suggests that the approximate quantum shell model works better for the 3-body systems (e.g., N baryons) than for 2-body systems such as mesons. This highly simplified model also works better for K mesons with larger masses than those with smaller masses. This property appears to suggest that light K mesons have a larger correlation between their two quarks than heavier K mesons.

To summarize, our model for the quantum 3-body problem of baryon formation is based on both a conceptual simplification and a simplification of the confining potential in quark Hamiltonians. The conceptual simplification is to reduce a 3-body collision to two consecutive 2-body collisions. This leads naturally to a quantum shell structure with a central quark moving in an effective harmonic oscillator potential, $V_{ho} = Qr^2 + V_o$. In addition, the confining potential $V = V_{ho}$ in the usual Dirac Hamiltonian, $H = \alpha_k p_k + \beta m_d + V$ is simplified to a special form

$$H_d \approx \alpha_k p_k + \beta m_d + \frac{(1+\beta)}{2} V, \qquad \text{(C66)}$$

that enables us to derive analytically the energy eigenvalue E_d in (C24).

The results in our model suggest three important new properties that could help to unveil the secret to predicting baryon masses:

(a) The general pattern of the energy (or mass) states of N baryons appears to be increasing with orbital angular momentum quantum number ℓ. This indicates that the effective harmonic oscillator potential must involve a function R_ℓ, which increases with the value of ℓ.

(b) The forces associated with the potentials V_{HO} and $V = Q'r$ are the same for all N baryon states consisting of the 'u u d' quarks. However, other baryon states such as the Σ baryon states that consist of 'u u s' quarks have different quantum shells. Consequently, their harmonic oscillator potential has a different (coupling) constant K, as shown in (C30) and (C31).

(c) The constant of integration V_0 (or $b = (E_d+m_d)V_0/(4a^2)$) in the C3Q model is interesting and strange because $b \approx 1$ is necessary to understand the spacings of the energy eigenstates of N baryons. (See (C25)–(C44).) However, the corresponding values of b' in the model for the K meson states do not seem to correspond. (See (C53)–(C66).)

In conclusion, the C3Q model seems to give a reasonable understanding of all 29 N baryon masses (equations (C25)–(C44)) based on two parameters K and b, together with an ℓ-dependent function R_ℓ in (C25). This appears to indicate that there may be some truth to the simplified Hamiltonians H_d in (C17) and H_u in (C3), the quantum shell suggested by the wave function $\Psi' \propto e^{-a^2 r^2/2}(a^2 r^2)^{(\ell+k)/2}$, and the permanent stability of the baryon ground state, i.e., the proton. The C3Q model with some modifications (e.g., to include additional isospin information) could be applied to understanding other baryon spectra, such as Σ baryons, Λ baryons, etc.[4] Furthermore, the C3Q model is also applicable to anti-baryons consisting of anti-quarks bound together with anti-color charges. It could help to understand the anti-N-baryons, their interactions and their early evolution in an antimatter half-universe.

References

1. J. P. Hsu, L. Hsu and D. Katz, Mod. Phys. Letts. A **33** 1850116 (2018).
2. J. P. Hsu and L. Hsu, *"Quantum Shell Model for Confining 3-Quark Baryons"*, UMassD preprint, (2023).
3. J. P. Hsu and L. Hsu, Chin. J. of Phys. **74**, 60-71 (2021); Ch. 3.
4. P. A. Zyla et al., (Particle Data Group) Prog. Theor. Exp. Phys., 2020, 083C01 (2020) pp. 91–102, pp. 312–324, see p. 312 for parity of K mesons.
5. E. Eichten, K. Gottfried, T. Kinoshita, J. Kogut, K. D. Lane, T.-Y. Yan, Phys. Rev. Lett., **34**, 369 (1975).
6. J. J. Sakurai, *Advanced Quantum Mechanics* (Addison-Wesley, 1967) pp. 22–125; R. Lisboa, M. Malheiro, A.S. de Castro, P. alberto and M. Fiolhais, Phys. Rev. C **69**, 024319 (2004). E. U. Condon and G. H. Shortly, *The Theory of Atomic Spectra* (Cambridge University Press, 1935).
7. N. Y. Sonine, Math. Ann. **16**, 1–80 (1880); M. E. Hassani, 'A note on Laguerre original ODEs and Polynomials (1879).' Google Scholar.

13

Epilogue

The framework of general Yang-Mills symmetry can encompass inertial and non-inertial frames, scales from the microscopic to the super-macroscopic, and all known interactions with enormous differences in their coupling strengths. It is the result of a consistent and fruitful collaboration.

Since our college years, two refreshingly innocent questions have become engrossed in our minds and guided our research and collaboration:

(i) Why can't one single time be consistent with and embedded in a four-dimensional framework?

(ii) Is the Lorentz transformation uniquely specified by precision experiments?

An answer to the first question took a definite step forward under a grapevine in a backyard garden in JP's freshman year. He suddenly saw a vivid picture in his mind's eye:

$$ct' \rightarrow c't$$

in the Lorentz transformations. Yes! It was possible to have a single time t in a four-dimensional transformation. However, everything involved with this refreshingly innocent change was very confusing, mainly due to deep-rooted preconceptions. Years later, it was realized that one should replace $c't$ with $b't$ in the transformations to avoid certain confusions. After fifteen years, it suddenly became trivial.... There was no mystery at all: One clock system could indeed be used by all observers in different inertial frames to record time of an event. However, the problem of convincing a referee of the validity of this idea remained. And so it was another twenty-one years until it met the eyes of a referee[a] for Foundations of Physics before it finally saw the light of publication.

In his junior year in college, Leon wrote a term paper for a seminar class that critically examined a claim in a Physical Review Letters paper that relativistic time was uniquely determined by precision experiments.[1] Introducing an arbitrary function for the evolution variable of a coordinate 4-vector, measured in units of length, he demonstrated that as long as this arbitrary function transformed as the fourth component of a 4-vector, precision experiments could not force an evolution variable, measured in units

---

[a]A friend of editor (W. Yourgrau) spent one week to ponder the idea. This concept was examined and later commented on by an editor of Nature in 1983.[2]

of seconds, to be relativistic time, unless a seemingly innocent assumption was made. This idea of an evolution variable with units of length had, of course, been explored previously by others.[3] However, its implications for the flexibility of the definition of time, measured in seconds, had not. The term paper served as a point of departure for the formulation of a broader relativity (called 'Taiji Relativity') based solely on the principle of relativity, without any postulate regarding the speed of light. This junior term paper also paved the way for the discovery of concrete space-time coordinate transformations between inertial and non-inertial frames, which were necessary for a broader understanding of the space-time properties of non-inertial frames and their operational meaning. Eventually, this understanding strengthened the conviction that it should be possible for all field theories, including gravity, to be formulated in inertial and non-inertial frames in flat space-time.

With this background, at the dawn of the 21st century, a final question emerged: 'How can gravity be understood in flat space-time?' 'I must find out the answer before I retire,' JP promised. There was basically only one approach to explore: The Yang-Mills approach to gravity based on space-time translational gauge symmetry in flat space-time. Following this approach strictly, JP encountered an enormously uncertainty. There are two independent quadratic terms in the T_4 gauge curvatures for the Lagrangian and both are completely different from the familiar linear curvature in the Hilbert-Einstein Lagrangian. Only a deep faith in the broader space-time of Taiji Relativity supported hundreds of pages of trial and error calculations. Ten years later, he finally convinced himself with an affirmative answer: Quantum Yang-Mills gravity could indeed be formulated in flat space-time and miraculously, also be consistent with experimental observations such as the perihelion shift of the Mercury, the deflection of light by gravity, and others.

What we have attempted here is to create a new perspective on dynamic symmetry and its role in a unified view of all basic interactions in the physical universe, from quark confinement to the late-time cosmic accelerated expansion to the creation of an antimatter half-universe with high energy electron-neutrinos as anti-darkmatter. For such a broad view of nature, it is natural to use Broader Particle-Cosmology as a base, since it is based on (i) general Yang-Mills symmetry, (ii) particle physics with quantum Yang-Mills gravity, and (iii) inertial and non-inertial frames of reference with space-time transformations among them. Importantly, the space and time coordinates of those frames have operational meanings.

General Yang-Mills symmetry gives a comprehensive overview of all basic interactions in physics. It reveals the fundamental principles of the physical universe:

(i) All internal gauge symmetries — $SU_3, (SU_2 \times U_1), U_{1b}, U_{1\ell}$ are only approximate symmetries due to their universal coupling to gravity. Consequently, electric charge, for example, is no longer exactly conserved and the electroweak theory is no longer renormalizable in the presence of gravity. Similarly, the usual QCD is not renormalizable and its asymptotic freedom is upset by the presence of quantum Yang-Mills gravity.

(ii) The space-time translational gauge symmetry of quantum Yang-Mills gravity is the only exact dynamical symmetry in physics — an important conclusion of the symmetry-unified model of all interactions. The geometric aspect of this translational symmetry is the exact conservation of the energy-momentum tensor.

Conventionally, it is believed that a field theory must be renormalizable in order to make sense. However, for all practical purposes, as long as a field theory can make predictions and is consistent with experiments, it should be satisfactory. Why must it have a finite number of counter terms?

References

1. D. Hils and J. L. Hall, Phys. Rev. Lett. **64**, 1697 (1990).
2. Editorial, Nature, **303**, 129 (1983).
3. See, for example, E. F. Taylor and J. A. Wheeler, *Spacetime Physics* (second ed. W. H. Freeman and Company, 1992) p. 60.

Author Index

Subject Index